Semiconductor Optoelectronic Devices

Introduction to Physics and Simulation

Semiconductor Optoelectronic Devices

Introduction to Physics and Simulation

JOACHIM PIPREK

University of California
at Santa Barbara

ACADEMIC PRESS

An Imprint of Elsevier

Amsterdam   Boston   London   New York   Oxford   Paris
San Diego   San Francisco   Singapore   Sydney   Tokyo

Academic Press
An Imprint of Elsevier
525 B Street, Suite 1900, San Diego, California 92101-4495, USA
http://www.academicpress.com

Academic Press
84 Theobald's Road, London WC1X 8RR, UK
http://www.academicpress.com

Library of Congress Control Number: 2002111026

ISBN-13: 978-0-12-557190-6
ISBN-10: 0-12-557190-9

To Lisa

Contents

Preface

Optoelectronics has become an important part of our lives. Wherever light is used to transmit information, tiny semiconductor devices are needed to transfer electrical current into optical signals and vice versa. Examples include light-emitting diodes in radios and other appliances, photodetectors in elevator doors and digital cameras, and laser diodes that transmit phone calls through glass fibers. Such optoelectronic devices take advantage of sophisticated interactions between electrons and light. Nanometer scale semiconductor structures are often at the heart of modern optoelectronic devices. Their shrinking size and increasing complexity make computer simulation an important tool for designing better devices that meet ever-rising performance requirements. The current need to apply advanced design software in optoelectronics follows the trend observed in the 1980s with simulation software for silicon devices. Today, software for technology computer-aided design (TCAD) and electronic design automation (EDA) represents a fundamental part of the silicon industry. In optoelectronics, advanced commercial device software has emerged, and it is expected to play an increasingly important role in the near future.

The target audience of this book is students, engineers, and researchers who are interested in using high-end software tools to design and analyze semiconductor optoelectronic devices. The first part of the book provides fundamental knowledge in semiconductor physics and in waveguide optics. Optoelectronics combines electronics and photonics and the book addresses readers approaching the field from either side. The text is written at an introductory level, requiring only a basic background in solid state physics and optics. Material properties and corresponding mathematical models are covered for a wide selection of semiconductors used in optoelectronics. The second part of the book investigates modern optoelectronic devices, including light-emitting diodes, edge-emitting lasers, vertical-cavity lasers, electroabsorption modulators, and a novel combination of amplifier and photodetector. InP-, GaAs-, and GaN-based devices are analyzed. The calibration of model parameters using available measurements is emphasized in order to obtain realistic results. These real-world simulation examples give new insight into device physics that is hard to gain without numerical modeling. Most simulations in this book employ the commercial software suite developed by Crosslight Software, Inc. (APSYS, LASTIP, PICS3D). Interested readers can obtain a free trial version of this software including example input files on the Internet at http://www.crosslight.com.

I would like to thank all my students in Germany, Sweden, Great Britain, Taiwan, Canada, and the United States, for their interest in this field and for all their questions, which eventually motivated me to write this book. I am grateful to Dr. Simon Li for creating the Crosslight software suite and for supporting my work. Prof. John Bowers gave me the opportunity to participate in several leading edge research projects, which provided some of the device examples in this book. I am also thankful to Prof. Shuji Nakamura for valuable discussions on the nitride devices. Parts of the manuscript have been reviewed by colleagues and friends, and I would like to acknowledge helpful comments from Dr. Justin Hodiak, Dr. Monica Hansen, Dr. Hans-Jürgen Wünsche, Daniel Lasaosa, Dr. Donato Pasquariello, and Dr. Lisa Chieffo. I appreciate especially the extensive suggestions I received from Dr. Hans Wenzel who carefully reviewed part I of the book.

Writing this book was part of my ongoing commitment to build bridges between theoretical and experimental research. I encourage readers to send comments by e-mail to piprek@ieee.org and I will continue to provide additional help and information at my web site http://www.engr.ucsb.edu/~piprek.

Joachim Piprek
Santa Barbara, California

List of Tables

Part I

Fundamentals

Chapter 1

Introduction to Semiconductors

This chapter gives a brief introduction to semiconductors. Electrons and holes
are carriers of electrical current in semiconductors and they are separated by
an energy gap. Photons are the smallest energy packets of light waves and
their interaction with electrons is the key physical mechanism in optoelectronic devices. The internal temperature of the semiconductor depends on the
energy of lattice vibrations, which can be divided into phonons. The Fermi
distribution function for the electron energy and the density of electron states
are introduced.

1.1 Electrons, Holes, Photons, and Phonons

Optoelectronics brings together optics and electronics within a single device, a single material. The material of choice needs to allow for the manipulation of light, the
manipulation of electrical current, and their interaction. Metals are excellent electrical conductors, but do not allow light to travel inside. Glass and related dielectric
materials can accommodate and guide light waves, like in optical fibers, but they
are electrical insulators. Semiconductors are in between these two material types, as
they can carry electrical current as well as light waves. Even better, semiconductors
can be designed to allow for the transformation of light into current and vice versa.

The conduction of electrical current is based on the flow of electrons. Most
electrons are attached to single atoms and are not able to move freely. Only some
loosely bound electrons are released and become conduction electrons. The same
number of positively charged atoms (ions) is left behind; the net charge is zero.
The positive charges can also move, as valence electrons jump from atom to
atom. Thus, both valence electrons (holes) and conduction electrons are able to
carry electrical current. Both the carriers are separated by an energy gap; i.e.,
valence electrons need to receive at least the gap energy E_g to become conduction
electrons. In semiconductors, the gap energy is on the order of 1 eV. The energy
can be provided, e.g., by light having a wavelength of less than the gap wavelength

$$\lambda_g = \frac{hc}{E_g} = \frac{1240\,\text{nm}}{E_g(\text{eV})} \tag{1.1}$$

3

with the light velocity c and Planck's constant h. In the wave picture, light is represented by periodic electromagnetic fields with the wavelength λ (see Chapter 4). In the particle picture, light is represented by a stream of energy packets (photons) with the energy

$$E_{\text{ph}} = \frac{hc}{\lambda} = h\nu = \hbar\omega \qquad (1.2)$$

(ν is frequency, $\omega = 2\pi\nu$ is angular frequency, and $\hbar = h/2\pi$ is the reduced Planck constant). The photon energy must be at least as large as the band gap E_{g} to generate electron–hole pairs. Vice versa, conduction electrons can also release energy in the form of light and become valence electrons. This energy exchange between electrons and photons is the key physical mechanism in optoelectronic devices.

From an atomic point of view, valence electrons belong to the outermost electron shell of the atom, which is fully occupied in the case of semiconductors; i.e., no more electrons with the same energy are allowed. As these atoms are joined together in a semiconductor crystal, the electrons start to interact and the valence energy levels separate slightly, forming a valence energy band (Fig. 1.1). Electrons within this band can exchange places but no charge flow is possible unless there is a hole. To generate holes, some electrons must be excited into the next higher energy band, the conduction band, which is initially empty. The concentration n of electrons in the conduction band and the concentration p of holes in the valence band control the electrical conductivity σ of semiconductors

$$\sigma = qn\mu_n + qp\mu_p \qquad (1.3)$$

with the elementary charge q and the mobility μ_n and μ_p of holes and electrons, respectively.

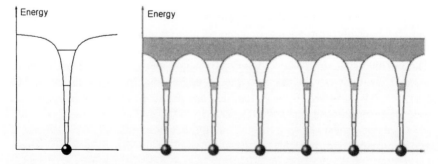

Figure 1.1: Electron energy levels of a single atom (left) become energy bands in a solid crystal (right).

Without external energy supply, the internal temperature T of the semiconductor governs the concentrations n and p. The higher the temperature, the stronger the vibration of the crystal lattice. According to the direction of the atom movement, those vibrations or lattice waves can be classified as follows:

- *longitudinal* (L) waves with atom oscillation in the travel direction of the lattice wave, and
- *transversal* (T) waves with atom oscillation normal to the travel direction.

According to the relative movement of neighbor atoms the lattice waves are

- *acoustic* (A) waves with neighbor atoms moving in the same direction, and
- *optical* (O) waves in ionic crystals with neighbor atoms moving in the opposite direction.

The last type of vibrations interacts directly with light waves as the electric field moves ions with different charges in different directions (see Chapter 4). Phonons represent the smallest energy portion of lattice vibrations, and they can be treated like particles. According to the classification above, four types of phonons are considered: LA, TA, LO, and TO. Electrons and holes can change their energy by generating or absorbing phonons.

1.2 Fermi Distribution and Density of States

The probability of finding an electron at an energy E is given by the Fermi distribution function

$$f(E) = \frac{1}{1 + \exp\left[\frac{E - E_F}{k_B T}\right]} \tag{1.4}$$

with the Fermi energy E_F and the Boltzmann constant k_B ($k_B T \approx 25$ meV at room temperature). At $T = 0$ K, the Fermi energy is the highest electron energy; i.e., it separates occupied from unoccupied energy levels. In pure semiconductors, E_F is typically somewhere in the middle of the band gap (Fig. 1.2). With increasing temperature, more and more electrons are transferred from the valence to the conduction band. The actual concentration of electrons and holes depends on the density of electron states $D(E)$ in both bands. Considering electrons and holes as (quasi-) free particles, the density of states in the conduction and valence band, respectively, becomes a parabolic function of the energy E (Fig. 1.2)

$$D_c(E) = \frac{1}{2\pi^2}\left(\frac{2m_c}{\hbar^2}\right)^{3/2}\sqrt{E - E_c} \quad (E > E_c) \tag{1.5}$$

$$D_v(E) = \frac{1}{2\pi^2}\left(\frac{2m_v}{\hbar^2}\right)^{3/2}\sqrt{E_v - E} \quad (E < E_v) \tag{1.6}$$

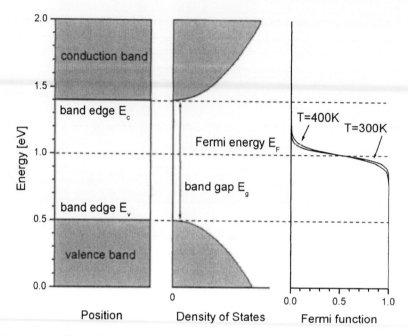

Figure 1.2: Illustration of energy bands, density of states, and Fermi distribution function.

with m_c and m_v being effective masses of electrons and holes, respectively. The carrier density as a function of energy is given by

$$n(E) = D_c(E)f(E) \tag{1.7}$$

$$p(E) = D_v(E)[1 - f(E)]. \tag{1.8}$$

Integration over the energy bands gives the total carrier concentrations

$$n \approx N_c \exp\left(\frac{E_F - E_c}{k_B T}\right) \tag{1.9}$$

$$p \approx N_v \exp\left(\frac{E_v - E_F}{k_B T}\right) \tag{1.10}$$

with the effective density of states

$$N_c = 2\left(\frac{m_c k_B T}{2\pi \hbar^2}\right)^{3/2} \tag{1.11}$$

$$N_v = 2\left(\frac{m_v k_B T}{2\pi \hbar^2}\right)^{3/2} \tag{1.12}$$

for the conduction and valence band, respectively. Equations (1.9) and (1.10) are valid for low carrier concentrations only ($n \ll N_c$, $p \ll N_v$); i.e., with the

Fermi energy separated from the band by more than $3k_BT$, allowing for the Boltzmann approximation

$$f(E) \approx f_B(E) = \exp\left[-\frac{E - E_F}{k_B T}\right].$$ (1.13)

If this condition is satisfied, like in pure (intrinsic) materials, the semiconductor is called nondegenerate. The intrinsic carrier concentration n_i is given as

$$n_i = \sqrt{np} = \sqrt{N_c N_v} \exp\left(-\frac{E_g}{2k_B T}\right).$$ (1.14)

At room temperature, n_i is very small in typical semiconductors, resulting in a poor electrical conductivity. Table 1.1 lists n_i and its underlying material parameters for various semiconductors.

Table 1.1: Energy Band Gap E_g, Density-of-States Effective Masses m_c and m_v, Effective Densities of States N_c and N_v, and Intrinsic Carrier Concentration n_i at Room Temperature [1, 2, 3, 4, 5, 6, 7]

Parameter	E_g	m_c	m_v	N_c	N_v	n_i
Unit	(eV)	(m_0)	(m_0)	(10^{18} cm^{-3})	(10^{18} cm^{-3})	(cm^{-3})
Si (X)	1.12	1.18	0.55	32.2	10.2	7×10^9
Ge (L)	0.66	0.22	0.34	2.6	5.0	1×10^{13}
GaAs (Γ)	1.42	0.063	0.52	0.40	9.41	2×10^6
InP (Γ)	1.34	0.079	0.60	0.56	11.6	1×10^7
AlAs (X)	2.15	0.79	0.80	17.6	18.1	15.
GaSb (Γ)	0.75	0.041	0.82	0.21	18.6	1×10^{12}
AlSb (X)	1.63	0.92	0.98	22.1	24.2	5×10^5
InAs (Γ)	0.36	0.023	0.57	0.09	10.9	9×10^{14}
GaP (X)	2.27	0.79	0.83	17.6	18.9	1.6
AlP (X)	2.45	0.83	0.70	20.0	14.8	0.044
InSb (Γ)	0.17	0.014	0.43	0.04	7.13	2×10^{16}
ZnS (Γ)	3.68	0.34	1.79	4.97	60.3	2×10^{-12}
ZnSe (Γ)	2.71	0.16	0.65	1.61	13.1	8×10^{-5}
CdS (Γ)	2.48	0.21	1.02	2.41	25.7	0.012
CdSe (Γ)	1.75	0.112	1.51	0.94	46.5	1×10^4
CdTe (Γ)	1.43	0.096	0.76	0.75	16.5	3×10^6

Note. Γ, direct semiconductor; X, L, indirect semiconductor; see Fig. 2.6. Parameters for GaN, AlN, and InN are given in Table 2.7.

1.3 Doping

To boost the concentration of electrons or holes, impurity atoms are introduced into the semiconductor crystal. As illustrated in Fig. 1.3, those dopants have energy levels slightly above the valence band (acceptors) or slightly below the conduction band (donors). Acceptors receive an additional electron from the valence band and become negatively charged ions, thereby generating a hole (p-doping). Donors release an electron into the conduction band and become positively charged ions (n-doping). Equation (1.14) is still valid in thermal equilibrium; however, the minority carrier concentration is now much smaller than the concentration of majority carriers. In other words, the Fermi level E_F is close to the majority carrier band edge (Fig. 1.4), and the Boltzmann approximation of Eqs. (1.9) and (1.10) is not valid any more (degenerate semiconductor). In Fermi statistics, the general expressions for the carrier concentrations are

$$n = N_c F_{1/2} \left(\frac{E_F - E_c}{k_B T} \right) \tag{1.15}$$

$$p = N_v F_{1/2} \left(\frac{E_v - E_F}{k_B T} \right), \tag{1.16}$$

where $F_{1/2}$ is the Fermi integral of order one-half, as obtained by integrating Eq. (1.7) or (1.8). Figure 1.5 plots Eq. (1.15) for GaAs as a function of $E_F - E_c$

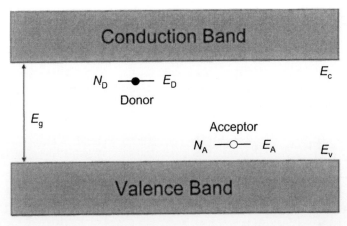

Figure 1.3: Illustration of donor and acceptor levels within the energy band gap (N_D, N_A, concentration; E_D, E_A, energy).

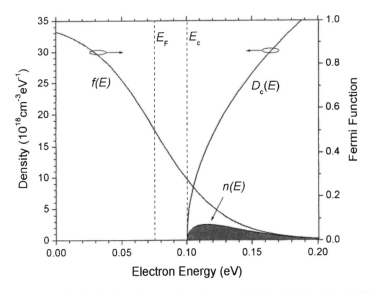

Figure 1.4: Parabolic density of energy band states $D(E)$ of GaAs (Eq. (1.5)) and Fermi distribution $f(E)$ with the Fermi level E_F slightly below the conduction band edge E_c. The gray area gives the carrier distribution $n(E)$ according to Eq. (1.7).

in comparison to the Boltzmann approximation (Eq. (1.9)). Increasing differences can be recognized as the Fermi level approaches the band edge. For numerical evaluation, the following approximation is often used for the Fermi integral and is indicated by the dots in Fig. 1.5 [8]:

$$F_{1/2}^{-1}(x) \approx e^{-x} + \frac{3}{4}\sqrt{\pi}\left\{x^4 + 50 + 33.6x\right.$$
$$\left. \times \left[1 - 0.68\exp(-0.17(x+1)^2)\right]\right\}^{-3/8}. \quad (1.17)$$

For bulk semiconductors in thermal equilibrium, the actual position of the Fermi level E_F is determined by the charge neutrality condition

$$p + p_D = n + n_A, \quad (1.18)$$

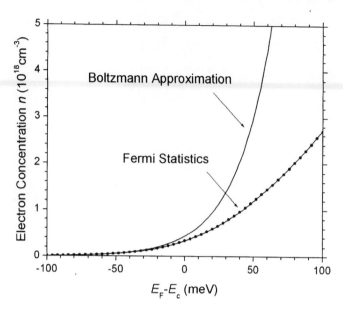

Figure 1.5: Electron concentration in the GaAs conduction band as a function of Fermi level position at room temperature. The result of the exact Fermi integral (Eq. (1.15)) is compared to the approximation by Eq. (1.17) (dots) and to the Boltzmann approximation (Eq. (1.9)).

where p_D is the concentration of ionized donors (charged positive) and n_A is the concentration of ionized acceptors (charged negative)

$$p_D = N_D \left\{ 1 + g_D \exp \left[\frac{E_F - E_D}{k_B T} \right] \right\}^{-1} \tag{1.19}$$

$$n_A = N_A \left\{ 1 + g_A \exp \left[\frac{E_A - E_F}{k_B T} \right] \right\}^{-1}. \tag{1.20}$$

Typical dopant degeneracy numbers are $g_D = 2$ and $g_A = 4$ [9]. E_D and E_A are the dopant energies (Fig. 1.3). Figure 1.6 plots the Fermi level position for n-doping or p-doping versus dopant concentration, as calculated from Eq. (1.18) for GaAs at room temperature. The Fermi level penetrates the conduction band with high n-doping and low ionization energy ($E_D = E_c - 0.01$ eV).

Nonequilibrium carrier distributions can be generated, for instance, by external carrier injection or by absorption of light. In such cases, electron and hole concentrations may be well above the equilibrium level. Each carrier distribution can still be characterized by Fermi functions, but with separate quasi-Fermi levels E_{Fn} and E_{Fp} for electrons and holes, respectively (see Section 3.2).

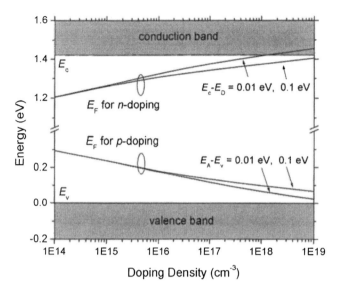

Figure 1.6: GaAs Fermi level E_F as a function of doping density for n-doping or p-doping with the ionization energy E_D and E_A as parameter ($T = 300\,\text{K}$).

Further Reading

- R. E. Hummel, *Electronic Properties of Materials*, 2nd ed., Springer-Verlag, Berlin, 1993.
- S. M. Sze, *Physics of Semiconductor Devices*, 2nd ed., Wiley, New York, 1981.

Chapter 2

Electron Energy Bands

This is a brief survey of important terms and theories related to the energy bands in semiconductors. First, the fundamental concepts of electron wave vectors $\vec{k}$, energy dispersion $E(\vec{k})$, and effective masses are introduced. Section 2.2 is mathematically more involved as it outlines the $\vec{k} \cdot \vec{p}$ method, which is most popular in optoelectronics for calculating the band structure. Semiconductor alloys, interfaces of different semiconductor materials, and quantum wells are covered at the end of this chapter.

2.1 Fundamentals

2.1.1 Electron Waves

In the classical picture, electrons are particles that follow Newton's laws of mechanics. They are characterized by their mass m_0, their position $\vec{r} = (x, y, z)$, and their velocity $\vec{v}$. However, this intuitive picture is not sufficient for describing the behavior of electrons within solid crystals, where it is more appropriate to consider electrons as waves. The wave–particle duality is one of the fundamental features of quantum mechanics. Using complex numbers, the wave function for a free electron can be written as

$$\psi(\vec{k}, \vec{r}) \propto \exp(i\, \vec{k}\vec{r}) = \cos(\vec{k}\vec{r}) + i \sin(\vec{k}\vec{r}) \tag{2.1}$$

with the wave vector $\vec{k} = (k_x, k_y, k_z)$. The wave vector is parallel to the electron momentum $\vec{p}$

$$\vec{k} = \frac{m_0 \vec{v}}{\hbar} = \frac{\vec{p}}{\hbar}, \tag{2.2}$$

and it relates to the electron energy E as

$$E - \frac{m_0}{2}v^2 = \frac{p^2}{2m_0} - \frac{\hbar^2 k^2}{2m_0}, \tag{2.3}$$

with $k^2 = k_x^2 + k_y^2 + k_z^2$. Hence, in all three directions, $E(\vec{p})$ and $E(\vec{k})$ are described by a parabola with the free electron mass m_0 as parameter.

Within semiconductors, an electron is exposed to the periodic lattice potential. It no longer behaves like a free particle as its de Broglie wavelength $2\pi/k$ comes close to the lattice constant a_0. The ensuing Bragg reflections prohibit a further acceleration of the electron, resulting in finite energy ranges for electrons, the energy bands.

In general, electron wave functions need to satisfy the Schrödinger equation

$$\frac{\hbar}{2m_0}\nabla^2\psi - V(\vec{r})\psi = E\psi, \tag{2.4}$$

where the potential $V(\vec{r})$ represents the periodic semiconductor crystal. This equation is often written as

$$H\psi = E\psi \tag{2.5}$$

with H called the Hamiltonian. The Schrödinger equation is for just one electron; all other electrons and atomic nuclei are included in the potential $V(\vec{r})$.[1] For the free electron, $V(\vec{r}) = 0$ and the solution to the Schrödinger equation is of the simple form given by Eq. (2.1). Within semiconductors, the solutions to the Schrödinger equation are so-called Bloch functions, which can be expressed as a linear combination of waves

$$\psi_n(\vec{k}, \vec{r}) = u_n(\vec{k}, \vec{r})\exp(i\vec{k}\vec{r}), \tag{2.6}$$

with the electron band index n. These functions are plane waves with a space-dependent amplitude factor $u_n(\vec{k}, \vec{r})$ that shows lattice periodicity. A one-dimensional schematic representation is given in Fig. 2.1 to indicate the relationship between the lattice potential and the Bloch function. The probability of finding an electron at the position $\vec{r}$ is proportional to $|\psi_n(\vec{r})|^2$.

In some practical cases, exact knowledge of the semiconductor Bloch functions is not required; only the energy dispersion function $E(\vec{k})$ needs to be found. Inserting Eq. (2.6) into Eq. (2.4), we obtain solutions only for certain ranges of the electron energy $E_n(\vec{k})$, the energy bands, which are separated by energy gaps (Fig. 2.2). A general feature of the solutions to the Schrödinger equation is the periodicity of $E_n(\vec{k})$, given in Fig. 2.2a. This figure shows the periodicity in the k_x direction with a period length of $k_x a_0 = 2\pi$; a shift of the solution $E(k_x)$ by $2\pi/a_0$ in k_x represents the same behavior. Any full segment of the periodic representation is a reduced k-vector representation. It is shown for the range $-\pi/a_0 < k_x < \pi/a_0$ in Fig. 2.2c. This k-range is called first Brillouin zone. The same treatment applies to the other two directions in the $\vec{k}$ space. Figure 2.3 illustrates the first Brillouin zone in two dimensions and it indicates several symmetry points. Besides the central Γ point, the zone boundaries exhibit additional symmetry points. The X point

---

[1]Many-body theories include the other particles explicitly [10].

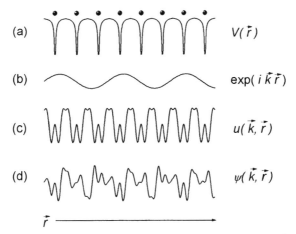

Figure 2.1: Schematic representation of electronic functions in a crystal: (a) potential plotted along a row of atoms, (b) free electron wave function, (c) amplitude factor of Bloch function having the periodicity of the lattice, and (d) Bloch function $\psi = u \exp(ikr)$.

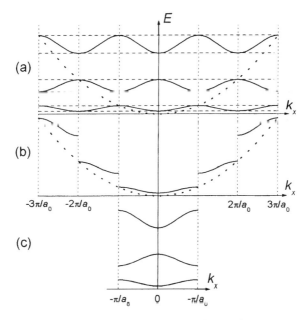

Figure 2.2: Comparison of different representations of the energy dispersion function $E(k)$: (a) periodic, (b) extended wave number, and (c) reduced wave number representation (a_0, lattice constant).

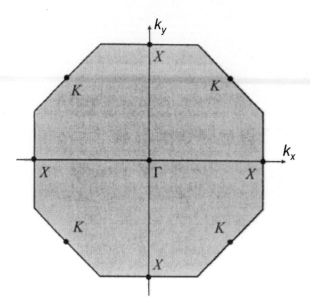

Figure 2.3: Two dimensional illustration of the first Brillouin zone for cubic (fcc) semiconductors with symmetry points Γ, X, and K. The X and K points lie in the $\langle 100 \rangle$ and $\langle 110 \rangle$ direction, respectively.

is shown in $\pm k_x$ and $\pm k_y$ directions, equivalent to the $\pm k_z$ directions, which are all denoted as the $\langle 100 \rangle$ direction. The K point lies in the $\langle 110 \rangle$ direction and the L point in the $\langle 111 \rangle$ direction. The electron band structure is fully described by the three-dimensional dispersion functions $E_n(\vec{k})$ within the first Brillouin zone. An example is given in Fig. 2.4 for GaAs exhibiting the smallest band gap at the Γ point.

The $\vec{k}$-space concept is explained in much more detail in many solid-state textbooks, e.g., in [12]. Brillouin zone properties like symmetry points depend on the crystal type. GaAs and most other common materials in optoelectronics are zinc-blende-type crystals (Fig. 2.5). Si and Ge form diamond-type crystals. Both types exhibit a so-called face-centered cubic (fcc) Bravais lattice. Some materials are able to form more than one crystal structure. For instance, GaN, AlN, and InN may exist as zinc blende crystals but they are commonly grown as wurtzite crystals, which exhibit a hexagonal Bravais lattice (cf. Fig. 2.10).

2.1.2 Effective Mass of Electrons and Holes

The properties of semiconductors are mainly determined by the behavior of electrons near the smallest band gap, where the relationship $E_n(\vec{k})$ is almost parabolic

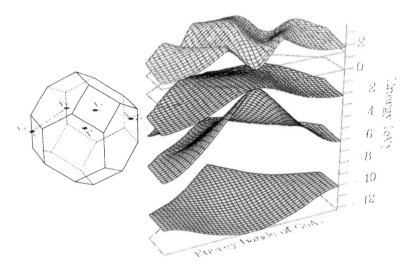

Figure 2.4: GaAs band structure with the smallest band gap at the Γ point [11].

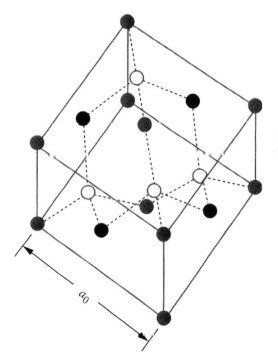

Figure 2.5: Zinc blende crystal with lattice constant a_0. The structure is formed by two intertwined face-centered cubic sublattices of, e.g., Ga and As atoms.

Table 2.1: Electron Effective Masses m_c in Units of m_0 for Conduction Band Minima in Cubic Semiconductors at Low Temperatures [2, 13]

Parameter	m_c^Γ	$m_c^X(t/l)$	$m_c^L(t/l)$
Si		0.191/0.916	
Ge	0.038		0.08/1.57
GaAs	0.067	0.23/1.3	0.0754/1.9
InP	0.0795	0.88(DOS)	0.47(DOS)
AlAs	0.15	0.22/0.97	0.15/1.32
GaSb	0.039	0.22/1.51	0.10/1.3
AlSb	0.14	0.123/1.357	0.23/1.64
InAs	0.026	0.16/1.13	0.05/0.64
GaP	0.13	0.253/2.0	0.15/1.2
AlP	0.22	0.155/2.68	
InSb	0.0135		0.25(DOS)

Note. Transversal/longitudinal electron masses are given for X and L points of the conduction band; DOS, density of states. Electron DOS masses at room temperature are given in Table 1.1 for the lowest band gap. For GaN, AlN, and InN, parameters for the common hexagonal crystal structure are given in Table 2.7.

so that electrons can be treated similar to free particles (cf. Eq. (2.3)). The free electron mass m_0 is replaced by an effective electron mass

$$m_n = \hbar^2 \left(\frac{d^2 E_n}{dk^2} \right)^{-1} \qquad (2.7)$$

to represent the influence of the crystal lattice. The effective electron mass m_n is inverse proportional to the curvature $d^2 E_n/dk^2$ of the energy dispersion. If $E_n(\vec{k})$ is anisotropic, different effective masses need to be given for each crystal direction (Table 2.1). The effective mass approximation allows for significant simplifications in semiconductor theory, and it is widely used in modeling optoelectronic devices. Effective mass calculations are based on approximate solutions to the Schrödinger equation. Examples are the tight binding, the orthogonalized plane wave, the pseudopotential, the cellular, the augmented plane wave, and the Green's function methods. These methods differ in the assumptions made for the crystal potential V and for the wave function ψ, and they are described in several textbooks (see, e.g., [12]).

In optoelectronic devices, electron transitions near the smallest gap between occupied (valence) and unoccupied (conduction) bands are of main interest. In typical optoelectronic materials based on, e.g., GaAs, InP, or GaN, this band gap occurs at the Γ point ($\vec{k} = 0$). In atomic physics terminology, the electron states at the bottom of the conduction band are s-like with zero orbital angular momentum (isotropic in space). The states at the top of the valence band are p-like with nonzero angular momentum and they are anisotropic (three independent states). Their orbital angular momentum is $\hbar$, whereas the spin angular momentum is $\hbar/2$. Both momenta can point in different directions and add up to different total momenta. The valence state notation often uses the total angular momentum and its value in the z direction (in $\hbar$):

$$\text{Heavy–Hole State} \quad \left|\tfrac{3}{2}, \pm\tfrac{3}{2}\right\rangle$$

$$\text{Light–Hole State} \quad \left|\tfrac{3}{2}, \pm\tfrac{1}{2}\right\rangle$$

$$\text{Split–Off State} \quad \left|\tfrac{1}{2}, \pm\tfrac{1}{2}\right\rangle.$$

In cubic semiconductors, heavy-hole (HH) and light-hole (LH) bands usually overlap at the Γ point (energy band degeneration). The spin–orbit (SO) split-off valence band is separated by the split-off energy Δ_0 and it is often less important. Figure 2.6 illustrates the bands of interest. Besides the direct band gap at the Γ point, additional conduction band minima may occur at other points in the first Brillouin zone. Figure 2.6 shows the case of an indirect semiconductor like Si where the lowest band gap is correlated to an indirect electron transition involving a change in momentum. However, most materials used in optoelectronics are direct semiconductors having the smallest band gap at the Γ point. For those materials, the most popular band-structure model is based on the $\vec{k} \cdot \vec{p}$ method, which is outlined in the following sections.

Electron effective masses are given in Table 2.1. The minima of the L and X conduction bands are typically anisotropic (ellipsoidal isoenergy surfaces) and they are characterized by a longitudinal and a transversal effective mass. The density-of-states (DOS) effective mass in Eq. (1.5) is then given by

$$m_c = \nu_D^{2/3} (m_l m_t^2)^{1/3} \tag{2.8}$$

with the degeneracy number ν_D of equivalent conduction band minima ($\nu_D = 6$ for Si; 4 for Ge; 3 for AlAs, AlSb, AlP). For example, the conduction band minimum of Si lies at the X point, which occurs $\nu_D = 6$ times in the first Brillouin zone (cf. Fig. 2.3). The longitudinal mass $m_l = 1.916 m_0$ in the $\langle 100 \rangle$ direction and the transversal mass $m_t = 0.191\ m_0$ for the two perpendicular directions was obtained from low-temperature cyclotron resonance measurements [14], resulting in the DOS mass $m_c = 1.064$. Due to temperature effects, this value is slightly lower than the room-temperature value $m_c = 1.18$ given in Table 1.1.

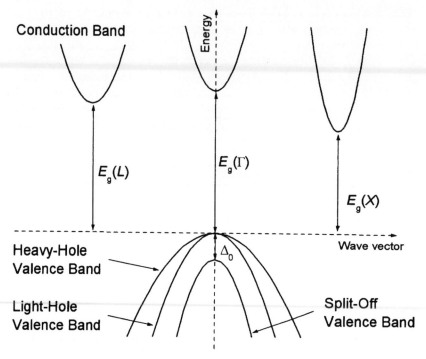

Figure 2.6: Illustration of bands and band gaps of interest.

Table 2.2 lists DOS effective masses for the three valence bands. The density of states near the Γ point is given by Eq. (1.6) with

$$m_v^{3/2} = m_{hh}^{3/2} + m_{lh}^{3/2}. \tag{2.9}$$

The effective mass typically increases with higher carrier energy, as the band curvature decreases, which makes the effective mass somewhat dependent on doping and on temperature [7]. The anisotropy of the valence bands and dependence of the hole effective masses on the crystal direction are discussed in Section 2.2.

2.1.3 Energy Band Gap

Energy gaps and related parameters are listed in Tables 1.1, 2.3, and 2.9 for zinc blende crystals and in Table 2.7 for wurtzite nitrides. Band gap reductions with higher temperature mainly arise from the change of the lattice constant. The following Varshni approximation [15] is often employed using empirical parameters

Table 2.2: Hole Effective Masses in Units of m_0 for the Heavy-Hole Band (m_{hh}), the Light-Hole Band (m_{lh}), and the Split-Off Band (m_{so}) at Room Temperature [1, 2, 4, 5, 6]

Parameter	m_{hh}	m_{lh}	m_{so}
Si	0.49	0.16	0.23
Ge	0.33	0.044	0.09
GaAs	0.50	0.076	0.14
InP	0.56	0.12	0.12
AlAs	0.76	0.15	0.24
GaSb	0.28	0.05	0.13
AlSb	0.94	0.14	0.29
InAs	0.57	0.025	0.14
GaP	0.79	0.14	0.46
AlP	0.63	0.20	0.29
InSb	0.43	0.015	0.19
ZnS	1.76	0.17	
ZnSe	0.6	0.15	
ZnTe	0.45	0.18	
CdS	0.64	0.64	
CdSe	>1	0.9	
CdTe	0.72	0.13	

A and B (Table 2.3)

$$E_g(T) = E_g(0) - \frac{AT^2}{B + T}. \tag{2.10}$$

A more accurate physics based formula has recently been proposed by Pässler [16]

$$E_g(T) = E_g(0) - \frac{\alpha\Theta}{2}\left[\sqrt[p]{1 + \left(\frac{2T}{\Theta}\right)^p} - 1\right] \tag{2.11}$$

using the slope parameter α, the average phonon temperature Θ, and the phonon dispersion parameter p, which have been obtained for a wide variety of materials (Table 2.4).

Table 2.3: Energy Band Gaps at $T = 0\,\text{K}$ and Varshni Parameters of Eq. (2.10) [2, 13, 15]

Parameter Unit	E_g^Γ (eV)	E_g^X (eV)	E_g^L (eV)	A^Γ (meV/K)	B^Γ (K)	A^X (meV/K)	B^X (K)	A^L (meV/K)	B^L (K)
Si	4.34	1.1557		0.391	125	0.7021	1108		
Ge	0.8893		0.741	0.6842	398			0.4561	210
GaAs	1.519	1.981	1.815	0.5405	204	0.46	204	0.605	204
InP	1.4236	2.384	2.014	0.363	162			0.363	162
AlAs	3.099	2.24	2.46	0.885	530	0.70	530	0.605	204
GaSb	0.812	1.141	0.875	0.417	140	0.475	94	0.597	140
AlSb	2.386	1.696	2.329	0.42	140	0.39	140	0.58	140
InAs	0.417	1.433	1.133	0.276	93	0.276	93	0.276	93
GaP	2.886	2.35	2.72			0.5771	372	0.5771	372
AlP	3.63	2.52	3.57	0.5771	372	0.318	588	0.318	588
InSb	0.235	0.63	0.93	0.32	170				

Note. For GaN, AlN, and InN, parameters for the common wurtzite crystal structure are given in Table 2.7. For GaP, $E_g^\Gamma = 2.78$ eV at 300 K.

High carrier concentrations in conduction and valence bands lead to band gap reduction (renormalization) due to enhanced carrier–carrier interaction [10]. The following universal fit formula can be derived from many-body theory ($N = n = p$) [17]:

$$\Delta E_g = -C \left[\frac{\varepsilon_{st}^5}{N} \left(m_0 \frac{m_c + m_v}{m_c m_v} + BT^2 \frac{\varepsilon_{st}}{N} \right) \right]^{-\frac{1}{4}} \qquad (2.12)$$

as a function of the static dielectric constant ε_{st}, the effective masses m_c and m_v, and the temperature T. The two fit parameters are $C = 3.9 \times 10^{-5}$ eV cm$^{3/4}$ and $B = 3.1 \times 10^{12}$ cm^{-3} K^{-2}. Results are plotted in Fig. 2.7 for several semiconductors.

Heavy doping also leads to band gap narrowing caused by carrier–carrier interaction as well as by the distortion of the crystal lattice (band tailing). Both band gap reduction mechanisms add up and they are often hard to separate. The following fit formula was obtained for Si as a function of the doping density N_{dop} (10^{18} cm^{-3}) [18]:[2]

$$\Delta E_g \, (\text{meV}) = -6.92 \left[\ln\left(\frac{N_{dop}}{0.13}\right) + \sqrt{\left(\ln\left(\frac{N_{dop}}{0.13}\right)\right)^2 + 0.5} \right] \qquad (2.13)$$

---

[2] A more general treatment is described in [19].

Table 2.4: Fundamental Energy Band Gap at $T = 0$ K (Type Given in Parentheses) and Pässler Parameters of Eq. (2.11) for Cubic Semiconductors [16]

Parameter	E_g	p	α	Θ
Unit	(eV)	(−)	(meV/K)	(K)
Si (X)	1.170	2.33	0.318	406
Ge (L)	0.744	2.38	0.407	230
GaAs (Γ)	1.519	2.44	0.472	230
InP (Γ)	1.424	2.51	0.391	243
AlAs (X)	2.229	2.32	0.362	218
GaSb (Γ)	0.811	2.57	0.375	176
AlSb (X)	1.686	1.90	0.343	226
InAs (Γ)	0.414	2.10	0.281	143
GaP (X)	2.339	2.09	0.480	358
AlP (X)	2.49	2.5	0.35	130
InSb (Γ)	0.234	2.68	0.250	136
ZnS (Γ)	3.841	2.76	0.532	240
ZnSe (Γ)	2.825	2.67	0.490	190
ZnTe (Γ)	2.394	2.71	0.454	145
CdS (Γ)	2.583	2.47	0.402	147
CdSe (Γ)	1.846	2.58	0.405	168
CdTe (Γ)	1.606	1.97	0.310	108

Note. Parameters for wurtzite GaN, AlN, and InN are given in Table 2.7.

2.2   Electronic Band Structure: The $\vec{k} \cdot \vec{p}$ Method

The most popular way to calculate the band structure $E(k)$ near the Γ point is the so-called $\vec{k} \cdot \vec{p}$ method. This model introduces the Bloch function (Eq. (2.6)) into the single-electron Schrödinger equation, resulting in

$$\left[-\frac{\vec{p} \cdot \vec{p}}{2m_0} + V(\vec{r}) + \frac{\hbar}{m_0} \vec{k} \cdot \vec{p} \right] u_n(\vec{k}, \vec{r}) = \left[E_n(\vec{k}) - \frac{\hbar^2 k^2}{2m_0} \right] u_n(\vec{k}, \vec{r}). \quad (2.14)$$

The momentum operator $\vec{p}$ stands for $-i\hbar\nabla$. The functions $u_n(\vec{k}, \vec{r})$ are constructed as linear expansions of the solutions $u_{n0} = u_n(0, \vec{r})$ at the Γ point

$$u_n(\vec{k}, \vec{r}) = \sum_m a_{nm} u_{m0}(\vec{r}). \quad (2.15)$$

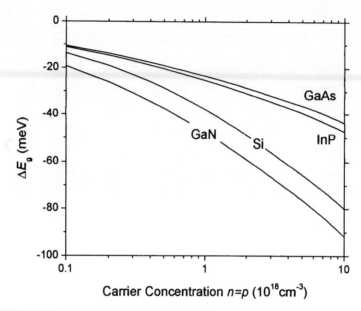

Figure 2.7: Band gap reduction as calculated from Eq. (2.12) at room temperature.

For small $\vec{k}$ vectors near the Γ point, perturbation theory can be applied as outlined in [20]. In first-order perturbation theory, the expansion coefficients a_{nm} are given in terms of a single material parameter, the bulk momentum matrix element M_b (cf. Section 5.1.1). This leads to Schrödinger matrix equations as large as twice the number of bands, including the two possible elecron spin directions.

2.2.1 Two-Band Model (Zinc Blende)

The conduction band is commonly well separated from the valence bands and treated as parabolic near the Γ point with the effective mass m_c. Even the spin–orbit split-off energy Δ_0 is often considered large enough to ignore the SO band (cf. Fig. 2.6). However, in GaAs- and InP-based materials, the other two valence bands are very close and cannot be treated separately. The interaction between HH and LH band leads to valence band mixing with nonparabolic dispersion functions $E(\vec{k})$. Including the spin degeneracy, the two-band $\vec{k} \cdot \vec{p}$ model gives a 4×4 Hamiltonian matrix, which is shown below together with the corresponding basis functions [21],

$$
\begin{bmatrix}
P_k + Q_k & -S_k & R_k & 0 \\
-S_k^* & P_k - Q_k & 0 & R_k \\
R_k^* & 0 & P_k - Q_k & S_k \\
0 & R_k^* & S_k^* & P_k + Q_k
\end{bmatrix}
\quad
\begin{matrix}
|3/2, +3/2\rangle \\
|3/2, +1/2\rangle \\
|3/2, -1/2\rangle \\
|3/2, -3/2\rangle
\end{matrix}
\qquad (2.16)
$$

with

$$P_k = \frac{\hbar^2 \gamma_1}{2m_0}(k_x^2 + k_y^2 + k_z^2) - E_v^0 \qquad (2.17)$$

$$Q_k = \frac{\hbar^2 \gamma_2}{2m_0}(k_x^2 + k_y^2 - 2k_z^2) \qquad (2.18)$$

$$R_k = -\sqrt{3}\frac{\hbar^2 \gamma_2}{2m_0}(k_x^2 - k_y^2) + i2\sqrt{3}\frac{\hbar^2 \gamma_3}{2m_0}k_x k_y \qquad (2.19)$$

$$S_k = 2\sqrt{3}\frac{\hbar^2 \gamma_3}{2m_0}(k_x - ik_y)k_z, \qquad (2.20)$$

where k_z is assumed to point in the $\langle 100 \rangle$ direction. E_v^0 marks the valence band edge; its absolute value is arbitrary, and only its offset between different semiconductors matters (cf. Table 2.9). R_k^* and S_k^* are the Hermitian conjugates of R_k and S_k, respectively, as generated by changing the sign in front of the imaginary unit i. The material constants γ_1, γ_2, and γ_3 are referred to as Luttinger parameters (Table 2.5). From the Hamiltonian matrix H_{ij} given by Eq. (2.16), the eigenvalues $E(\vec{k})$ can be found by solving the determinantal equation

$$\det\left[H_{ij} - \delta_{ij}E\right] = 0 \qquad (2.21)$$

Table 2.5: Luttinger Parameters γ for Cubic Semiconductors at Low Temperatures [2, 13]

Parameter	γ_1	γ_2	γ_3
Si	4.285	0.339	1.446
Ge	13.38	4.28	5.68
GaAs	6.98	2.06	2.93
InP	5.08	1.60	2.10
AlAs	3.76	0.82	1.42
GaSb	13.4	4.7	6.0
AlSb	5.18	1.19	1.97
InAs	20.0	8.5	9.2
GaP	4.05	0.49	2.93
AlP	3.35	0.71	1.23
InSb	34.8	15.5	16.5
ZnSe	4.3	1.14	1.84
ZnTe	3.9	0.83	1.30
CdTe	5.3	1.7	2.0

using the Kronecker symbol δ_{ij} ($\delta_{ij} = 1$ for $i = j$, $\delta_{ij} = 0$ otherwise). Without the nondiagonal elements in the Hamiltonian matrix H_{ij} ($S_k = R_k = 0$), simple parabolic dispersion functions $E(\vec{k})$ would result from Eq. (2.21) and the four bands would be independent of each other. The nondiagonal elements S_k and R_k lead to valence band mixing; i.e., each band's structure $E_n(\vec{k})$ is affected by the other bands. By changing the set of basis functions, the 4×4 Hamiltonian matrix can be transformed into two 2×2 Hamiltonians and Eq. (2.21) can be solved analytically [21]. This results in the energy dispersion functions

$$E_{\text{hh}}(\vec{k}) = P_k + \sqrt{Q_k^2 + |R_k|^2 + |S_k|^2} \tag{2.22}$$

$$E_{\text{lh}}(\vec{k}) = P_k - \sqrt{Q_k^2 + |R_k|^2 + |S_k|^2}, \tag{2.23}$$

giving

$$E(\vec{k}) = E_{\text{v}}^0 - \frac{\hbar^2}{2m_0}\left[\gamma_1 k^2 \pm \sqrt{4\gamma_2^2 k^4 + 12(\gamma_3^2 - \gamma_2^2)(k_x^2 k_y^2 + k_x^2 k_z^2 + k_y^2 k_z^2)}\right]. \tag{2.24}$$

Figure 2.8 illustrates the anisotropic nature of this solution. It is parabolic in each direction in $\vec{k}$ space, and the effective masses for the three main directions are

$$m_{\text{hh}}^{\langle 100 \rangle} = \frac{m_0}{\gamma_1 - 2\gamma_2} \tag{2.25}$$

$$m_{\text{lh}}^{\langle 100 \rangle} = \frac{m_0}{\gamma_1 + 2\gamma_2} \tag{2.26}$$

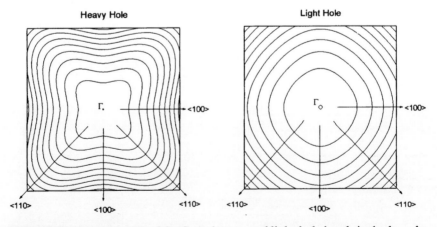

Figure 2.8: Isoenergy plot of the GaAs heavy- and light-hole bands in the $k_x - k_z$ plane near the Γ point.

$$m_{hh}^{(110)} = \frac{2m_0}{2\gamma_1 - \gamma_2 - 3\gamma_3} \qquad (2.27)$$

$$m_{lh}^{(110)} = \frac{2m_0}{2\gamma_1 + \gamma_2 + 3\gamma_3} \qquad (2.28)$$

$$m_{hh}^{(111)} = \frac{m_0}{\gamma_1 - 2\gamma_3} \qquad (2.29)$$

$$m_{lh}^{(111)} = \frac{m_0}{\gamma_1 + 2\gamma_3}. \qquad (2.30)$$

For small anisotropy, the axial approximation $\gamma_3^* = \gamma_2^*$ can be used, which gives spherical isoenergy surfaces with isotropic effective masses

$$m_{hh} = \frac{m_0}{\gamma_1 - 2\gamma_2^*} \qquad (2.31)$$

$$m_{lh} = \frac{m_0}{\gamma_1 + 2\gamma_2^*}. \qquad (2.32)$$

Different formulas for the averaged parameter γ_2^* can be found in the literature, for example [20],

$$\gamma_2^* = \frac{\gamma_2 + \gamma_3}{2}. \qquad (2.33)$$

2.2.2 Strain Effects (Zinc Blende)

Strain deforms the crystal lattice. However, it is assumed to be still periodic such that Bloch functions are still applicable. Since the elementary crystal cell is deformed, potential and Bloch lattice functions $u_n^s(\vec{k}, \vec{r})$ now have a period equal to the strained elementary cell. The relative change of the lattice period gives the strain

$$\varepsilon_{ij} = \frac{\Delta a_i}{a_j}, \qquad (2.34)$$

which may be different in each direction (a, lattice constant; $i, j = x, y,$ or z). The fractional change in volume is given by

$$\frac{\Delta V}{V} = \varepsilon_{xx} + \varepsilon_{yy} + \varepsilon_{zz}. \qquad (2.35)$$

With small strain, the strained Hamiltonian can be given as perturbation of the unstrained Hamiltonian H_0 [22]:

$$\left[\mathbf{H_0} + \mathbf{H_\varepsilon} \right] u_n^s(\vec{k}, \vec{r}) = E_n(\vec{k}) u_n^s(\vec{k}, \vec{r}). \qquad (2.36)$$

In the following, we consider the common case of biaxial strain, like in lattice mismatched superlattices or quantum wells grown in the z direction, with

$\varepsilon_{xx} = \varepsilon_{yy} \neq \varepsilon_{zz}$ and $\varepsilon_{ij} = 0$ for $i \neq j$. The two strain components are related by the elastic stiffness constants C_{11} and C_{12} as

$$\varepsilon_{zz} = -2\frac{C_{12}}{C_{11}}\varepsilon_{xx} \text{ with } \varepsilon_{xx} = \varepsilon_{yy} = \frac{a_{st} - a_0}{a_0}, \qquad (2.37)$$

where a_{st} and a_0 are the lattice constant of the strained and unstrained crystal, respectively. For compressive strain, $a_{st} < a_0$, $\varepsilon_{xx} = \varepsilon_{yy} < 0$, and $\varepsilon_{zz} > 0$. Additions P_ε and Q_ε are required to the diagonal elements of the two-band Hamiltonian matrix in Eq. (2.16) ($P_k \rightarrow P_k + P_\varepsilon$, $Q_k \rightarrow Q_k + Q_\varepsilon$) with

$$P_\varepsilon = -a_v(\varepsilon_{xx} + \varepsilon_{yy} + \varepsilon_{zz}) \qquad (2.38)$$

$$Q_\varepsilon = -\frac{b}{2}(\varepsilon_{xx} + \varepsilon_{yy} - 2\varepsilon_{zz}). \qquad (2.39)$$

As a result of this modification, the strain is found to cause the following shifts of the band edges at the Γ point:

$$E_{hh}(0) = E_{hh}^0 = E_v^0 - P_\varepsilon - Q_\varepsilon \qquad (2.40)$$

$$E_{lh}(0) = E_{lh}^0 = E_v^0 - P_\varepsilon + Q_\varepsilon. \qquad (2.41)$$

The conduction band edge is given by

$$E_c(0) = E_c^0 = E_v^0 + E_g + a_c(\varepsilon_{xx} + \varepsilon_{yy} + \varepsilon_{zz}). \qquad (2.42)$$

The factors a_c and a_v are so-called hydrostatic deformation potentials; b is the shear deformation potential (Table 2.6). The separation of the total hydrostatic deformation potential in conduction (a_c) and valence band (a_v) contributions is important at heterointerfaces (see Section 2.5).

The energy dispersion becomes

$$E_{hh}(\vec{k}) = E_v^0 - P_k - P_\varepsilon - \text{sgn}(Q_\varepsilon)\sqrt{(Q_k + Q_\varepsilon)^2 + |R_k|^2 + |S_k|^2} \qquad (2.43)$$

$$E_{lh}(\vec{k}) = E_v^0 - P_k - P_\varepsilon + \text{sgn}(Q_\varepsilon)\sqrt{(Q_k + Q_\varepsilon)^2 + |R_k|^2 + |S_k|^2}. \qquad (2.44)$$

The factor $\text{sgn}(Q_\varepsilon)$ is negative for compressive strain ($Q_\varepsilon < 0$) and positive for tensile strain ($Q_\varepsilon > 0$). Similar to Eq. (2.24), we obtain for the strained case and $k_y = 0$ [21]

$$E(k_x, 0, k_z) = E_v^0 - P_\varepsilon - \frac{\hbar^2}{2m_0}\left[\gamma_1(k_x^2 + k_z^2) \right.$$

$$\left. \pm\sqrt{\left(\gamma_2(k_x^2 - 2k_z^2) + \frac{2m_0}{\hbar^2}Q_\varepsilon\right)^2 + 3\gamma_2^2 k_x^4 + 12\gamma_3^2 k_x^2 k_z^2} \right]. \qquad (2.45)$$

Table 2.6: Lattice Constant a_0, Thermal Expansion Coefficient da_0/dT, Elastic Stiffness Constants C_{11} and C_{12}, and Deformation Potentials b, a_v, a_c for Cubic Semiconductors at Room Temperature [1, 2, 13, 23]

Parameter Unit	a_0 (Å)	da_0/dT $(10^{-5}$ Å/K)	C_{11} (GPa)	C_{12} (GPa)	b (eV)	a_v (eV)	a_c (eV)
Si	5.43102	1.41	16.577	6.393	−2.1	2.46	+4.18 (X)
Ge	5.6579	3.34	12.40	4.13	−2.9	1.24	−1.54 (L)
GaAs	5.65325	3.88	1221	566	−2.0	1.16	−7.17
InP	5.8697	2.79	1011	561	−2.0	0.6	−6.0
AlAs	5.6611	2.90	1250	534	−2.3	2.47	−5.64
GaSb	6.0959	4.72	884.2	402.6	−2.0	0.8	−7.5
AlSb	6.1355	2.60	876.9	434.1	−1.35	1.4	−4.5
InAs	6.0583	2.74	832.9	452.6	−1.8	1.00	−5.08
GaP	5.4505	2.92	1405	620.3	−1.6	1.7	−8.2
AlP	5.4672	2.92	1330	630	−1.5	3.0	−5.7
InSb	6.4794	3.48	684.7	373.5	−2.0	0.36	−6.94
ZnS	5.4102	3.68	9.81	6.27	−0.8	2.31	−4.09
ZnSe	5.6676	3.91	9.009	5.34	−1.2	1.65	−4.17
ZnTe	6.1037	5.00	7.13	4.07		0.79	−4.09
CdS	5.818		8.31	5.04			
CdSe	4.2999		7.41	4.52			
CdTe	6.486	3.11	5.33	3.65	−1.2	0.55	−3.96

Note. The deformation potential a_c is given for the lowest conduction band minimum at the Γ point unless indicated otherwise. For parameters of hexagonal compounds, see Table 2.7.

The band structure is illustrated in Fig. 2.9. In the k_z direction, the bulk bands are still parabolic with effective masses identical to the unstrained case

$$m_{hh}^z = \frac{m_0}{\gamma_1 - 2\gamma_2} \qquad (2.46)$$

$$m_{lh}^z = \frac{m_0}{\gamma_1 + 2\gamma_2}. \qquad (2.47)$$

However, within the k_x–k_y plane, the bands become nonparabolic and the effective masses depend on the wave number. Near the Γ point, the transversal

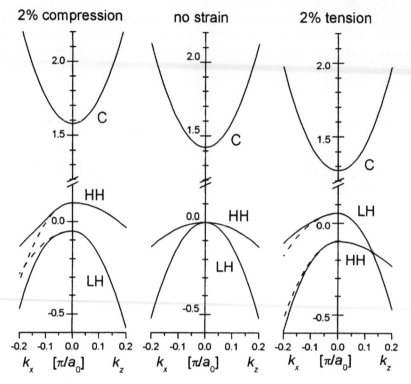

Figure 2.9: Shift and deformation of GaAs energy bands for compressive, zero, and tensile strain (C, conduction band; HH, heavy hole band; LH, light hole band). The vertical axes give the energy in electron volts, the lateral axes give the wave number k_x and k_z, respectively, within 20% of the first Brillouin zone. Dashed lines give the effective mass approximation (Eqs. (2.48), (2.49)).

effective masses are

$$m_{hh}^t = \frac{m_0}{\gamma_1 + \gamma_2} \tag{2.48}$$

$$m_{lh}^t = \frac{m_0}{\gamma_1 - \gamma_2}. \tag{2.49}$$

2.2.3   Three- and Four-Band Models (Zinc Blende)

All of the above results for heavy- and ligh-hole bands have neglected their interaction with the spin–orbit band as well as with the conduction band. The inclusion of such interactions is briefly described in the following.

The 6×6 $\vec{k} \cdot \vec{p}$ model solves the Schrödinger equation for the six valence bands, including spin degeneracy. In the Luttinger–Kohn model [24], the expansion of the functions $u_n(\vec{k}, \vec{r})$ also considers contributions from basis functions u_{n0} of higher bands. Including biaxial strain, the Hamiltonian matrix becomes for bulk material [20]

$$
\begin{bmatrix}
P+Q & -S & R & 0 & -\frac{S}{\sqrt{2}} & \sqrt{2}R \\
-S^* & P-Q & 0 & R & -\sqrt{2}Q & \sqrt{1.5}S \\
R^* & 0 & P-Q & S & \sqrt{1.5}S^* & \sqrt{2}Q \\
0 & R^* & S^* & P+Q & -\sqrt{2}R^* & -\frac{S^*}{\sqrt{2}} \\
-\frac{S^*}{\sqrt{2}} & -\sqrt{2}Q & \sqrt{1.5}S & -\sqrt{2}R & P+\Delta_0 & 0 \\
\sqrt{2}R^* & \sqrt{1.5}S^* & \sqrt{2}Q & -\frac{S}{\sqrt{2}} & 0 & P+\Delta_0
\end{bmatrix}
\begin{matrix}
|3/2, +3/2\rangle \\
|3/2, +1/2\rangle \\
|3/2, -1/2\rangle \\
|3/2, -3/2\rangle \\
|1/2, +1/2\rangle \\
|1/2, -1/2\rangle
\end{matrix}
\quad (2.50)
$$

with $P = P_k + P_\varepsilon$, $Q = Q_k + Q_\varepsilon$, $R = R_k$, and $S = S_k$ as given above ($S_\varepsilon = R_\varepsilon = 0$ for biaxial strain). The strain results in the following shifts of the band edges at the Γ point

$$E_{hh}^0 = E_v^0 - P_\varepsilon - Q_\varepsilon \tag{2.51}$$

$$E_{lh}^0 = E_v^0 - P_\varepsilon + \frac{1}{2}\left[Q_\varepsilon - \Delta_0 + \sqrt{\Delta_0^2 + 9Q_\varepsilon^2 + 2Q_\varepsilon\Delta_0}\right] \tag{2.52}$$

$$E_{so}^0 = E_v^0 - P_\varepsilon + \frac{1}{2}\left[Q_\varepsilon - \Delta_0 - \sqrt{\Delta_0^2 + 9Q_\varepsilon^2 + 2Q_\varepsilon\Delta_0}\right] \tag{2.53}$$

which, for the light-hole band, is somewhat different from the 4×4 Hamiltonian result in Eq. (2.41) due to the interaction with the SO band. The conduction band edge is still given by Eq. (2.12). Analytical formulas for the band structure $E(\vec{k})$ are hard to obtain in this case; for details see [25]. From the series expansion of E up to the second order of k near the Γ point, the effective masses for $k = 0$ can be extracted as

$$m_{hh}^z = \frac{m_0}{\gamma_1 - 2\gamma_2} \tag{2.54}$$

$$m_{hh}^t = \frac{m_0}{\gamma_1 + \gamma_2} \tag{2.55}$$

$$m_{lh}^z = \frac{m_0}{\gamma_1 + 2\gamma_2 f_+} \tag{2.56}$$

$$m_{lh}^t = \frac{m_0}{\gamma_1 - \gamma_2 f_+} \tag{2.57}$$

$$m_{so}^z = \frac{m_0}{\gamma_1 + 2\gamma_2 f_-} \tag{2.58}$$

$$m_{so}^t = \frac{m_0}{\gamma_1 - \gamma_2 f_-} \tag{2.59}$$

employing the strain factor

$$f_{\pm}(s) = \frac{2s[1 + 1.5(s - 1 \pm \sqrt{1 + 2s + 9s^2})] + 6s^2}{0.75(s - 1 \pm \sqrt{1 + 2s + 9s^2})^2 + s - 1 \pm \sqrt{1 + 2s + 9s^2} - 3s^2}$$

(2.60)

with the strain parameter $s = Q_\varepsilon / \Delta_0$. Without strain, $s = 0$ and $f_+(0) = 1$, the effective masses for heavy and light holes are identical to those obtained from the 4×4 Hamiltonian.

An 8×8 Hamiltonian matrix emerges if the conduction band is also included in the set of basis functions [26]. This four-band $\vec{k} \cdot \vec{p}$ model requires much algebra and the interested reader is referred to the literature (see, e.g., [27] or [28]). In most practical cases, conduction band and valence bands can be treated separately as shown above. However, one important result relates the electron effective mass m_c to the bulk momentum matrix element M_b

$$m_c = \left(\frac{1 + 2F_b}{m_0} + \frac{2M_b^2}{m_0^2} \frac{(3E_g + 2\Delta_0)}{E_g(E_g + \Delta_0)} \right)^{-1},$$

(2.61)

which shall be further employed in Chap. 5. F_b is a correction parameter that accounts for the influence of higher bands, which tend to make the effective mass heavier. For example, the electron effective mass $m_c/m_0 = 0.067$ of GaAs is reduced to 0.05 with $F_b = 0$. Both M_b and F_b are hardly temperature dependent so that the temperature effect on m_c mainly originates from the band gap E_g. The momentum matrix element is often given in terms of the energy parameter E_p (cf. Table 5.1)

$$M_b^2 = \frac{m_0}{6} E_p.$$

(2.62)

2.2.4 Three-Band Model for Wurtzite Crystals

Due to the recent success with GaN-based optoelectronic devices, their wurtzite (hexagonal) crystal system has gained much interest (cf. Chap. 9). The previous sections have outlined the $\vec{k} \cdot \vec{p}$ band-structure model for zinc blende (cubic) crystals. Here, we apply the $\vec{k} \cdot \vec{p}$ method in a similar way to wurtzite crystals. The wurtzite crystal structure is illustrated in Fig. 2.10. For wurtzite structures, all three valence bands are strongly coupled and valence band mixing must be taken into account. The three valence bands are referred to as heavy-hole (HH), light-hole (LH), and crystal-field split-hole (CH) band. Spin–orbit interaction leads to a slight separation of all three band edges. Wurtzite band-structure parameters are listed in Table 2.7. For GaN, Δ_{cr} is positive and E_{hh}^0 is the top valence band edge.

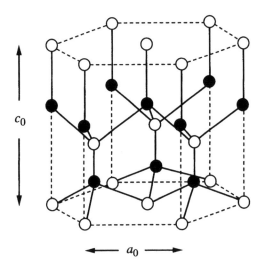

Figure 2.10: Wurtzite crystal with lattice constants c_0 and a_0. The structure is formed by two intertwined hexagonal sublattices of, e.g., Ga and N atoms.

For AlN, Δ_{cr} is negative and E_{lh}^0 is the top valence band edge. The band lineup at the GaN/AlN interface is shown in Fig. 2.11. The CH energy without spin–orbit interaction is used as reference energy E_v^0 ($\Delta_{so} = 0$).

In typical growth direction along the c axis of the wurtzite crystal, a strained layer with lattice constant $a_{st} = a_s$ is described by a diagonal strain tensor (a_s, substrate lattice constant; a_0, lattice constant of unstrained layer; $c \parallel z$)

$$\varepsilon_{xx} = \varepsilon_{yy} = \frac{a_s - a_0}{a_0}$$

$$\varepsilon_{zz} = -2\frac{C_{13}}{C_{33}}\varepsilon_{xx} \qquad (2.63)$$

$$\varepsilon_{xy} = \varepsilon_{yz} = \varepsilon_{zx} = 0.$$

Considering strain effects, the valence band edge energies are [29]

$$E_{hh}^0 = E_v^0 + \Delta_1 + \Delta_2 + \theta_\varepsilon + \lambda_\varepsilon \qquad (2.64)$$

$$E_{lh}^0 = E_v^0 + \frac{\Delta_1 - \Delta_2 + \theta_\varepsilon}{2} + \lambda_\varepsilon + \sqrt{\left(\frac{\Delta_1 - \Delta_2 + \theta_\varepsilon}{2}\right)^2 + 2\Delta_3^2} \qquad (2.65)$$

$$E_{ch}^0 = E_v^0 + \frac{\Delta_1 - \Delta_2 + \theta_\varepsilon}{2} + \lambda_\varepsilon - \sqrt{\left(\frac{\Delta_1 - \Delta_2 + \theta_\varepsilon}{2}\right)^2 + 2\Delta_3^2} \qquad (2.66)$$

Table 2.7: Electron Band-Structure Parameters for Nitride Wurtzite Semiconductors at Room Temperature [13, 16, 29, 31, 32, 33, 34, 35, 36, 37, 38]

Parameter	Symbol	Unit	InN	GaN	AlN
Electron eff. mass (c axis)	m_c^z	m_0	0.11	0.20	0.33
Electron eff. mass (transversal)	m_c^t	m_0	0.11	0.18	0.25
Momentum matrix parameter	E_p	eV	14.6	14.0	14.5
Hole eff. mass parameter	A_1	—	−9.24	−7.24	−3.95
Hole eff. mass parameter	A_2	—	−0.60	−0.51	−0.27
Hole eff. mass parameter	A_3	—	8.68	6.73	3.68
Hole eff. mass parameter	A_4	—	−4.34	−3.36	−1.84
Hole eff. mass parameter	A_5	—	−4.32	−3.35	−1.92
Hole eff. mass parameter	A_6	—	−6.08	−4.72	−2.91
Valence band reference level	E_v^0	eV	−1.59	−2.64	−3.44
Direct band gap (300 K)	E_g	eV	1.89	3.42	6.28
Direct band gap (0 K)	E_g	eV	1.994	3.47	6.20
Varshni parameter	A	meV/K	0.245	0.909	1.799
Varshni parameter	B	K	624	830	1462
Pässler Parameter	p	—	2.9	2.62	3.0
Pässler Parameter	Θ	K	453	504	575
Pässler Parameter	α	meV/K	0.21	0.599	0.83
Spin–orbit split energy	Δ_{so}	eV	0.001	0.014	0.019
Crystal-field split energy	Δ_{cr}	eV	0.041	0.019	−0.164
Lattice constant	a_0	Å	3.548	3.189	3.112
Lattice constant	c_0	Å	5.703	5.185	4.982
Elastic constant	C_{33}	GPa	200	392	382
Elastic constant	C_{13}	GPa	94	100	127
Hydrost. deform. potential (E_c)	a_c	eV		−4.08	
Shear deform. potential	D_1	eV		−0.89	
Shear deform. potential	D_2	eV		4.27	
Shear deform. potential	D_3	eV		5.18	
Shear deform. potential	D_4	eV		−2.59	
LO phonon energy	$\hbar\omega_{LO}$	meV	73	91	110

Note. $\Delta_{cr} = \Delta_1$, $\Delta_{so} = 3\Delta_2 = 3\Delta_3$, $a_c = a/2$.

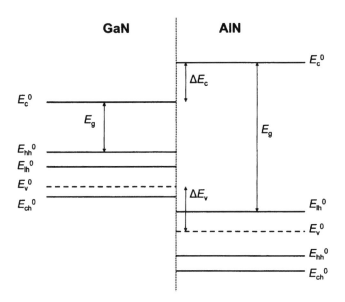

Figure 2.11: Relative position of band edge energies in GaN ($\Delta_{cr} > 0$) and AlN ($\Delta_{cr} < 0$), after [29].

with

$$\theta_\varepsilon = D_3\varepsilon_{zz} + D_4(\varepsilon_{xx} + \varepsilon_{yy}) \tag{2.67}$$

$$\lambda_\varepsilon = D_1\varepsilon_{zz} + D_2(\varepsilon_{xx} + \varepsilon_{yy}). \tag{2.68}$$

The valence band deformation potentials D_1 to D_4 are listed in Table 2.7. The conduction band exhibits a hydrostatic energy shift $P_{c\varepsilon}$

$$E_c^0 = E_v^0 + \Delta_1 + \Delta_2 + E_g + P_{c\varepsilon} \tag{2.69}$$

with

$$P_{c\varepsilon} = a_{cz}\varepsilon_{zz} + a_{ct}(\varepsilon_{xx} + \varepsilon_{yy}). \tag{2.70}$$

The hydrostatic deformation potentials are often assumed to be equal $a_c = a_{ct} = a_{cz}$. The GaN band edge shift due to biaxial compressive strain is plotted in Fig. 2.12.

The dispersion $E_c(\vec{k})$ of the conduction band can be characterized by a parabolic band model with electron effective masses m_c^t and m_c^z perpendicular and parallel to the c-growth direction, respectively (Table 2.7). The strong coupling of the three valence bands requires the use of a three band 6×6 Hamiltonian

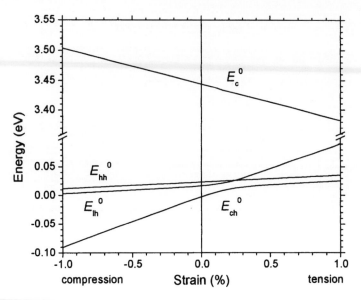

Figure 2.12: GaN band edge shift vs biaxial strain for conduction band and valence bands.

matrix, which can be block-diagonalized, giving two 3×3 Hamiltonians [29, 30]

$$H_{6\times6}^{v}(k) = \begin{bmatrix} H_{3\times3}^{U}(k) & 0 \\ 0 & H_{3\times3}^{L}(k) \end{bmatrix}, \qquad (2.71)$$

where

$$H^{U} = \begin{bmatrix} F & K_t & -iH_t \\ K_t & G & \Delta - iH_t \\ iH_t & \Delta + iH_t & \lambda \end{bmatrix} \qquad (2.72)$$

$$H^{L} = \begin{bmatrix} F & K_t & iH_t \\ K_t & G & \Delta + iH_t \\ -iH_t & \Delta - iH_t & \lambda \end{bmatrix} \qquad (2.73)$$

with

$$F = \Delta_1 + \Delta_2 + \lambda + \theta \qquad (2.74)$$
$$G = \Delta_1 - \Delta_2 + \lambda + \theta \qquad (2.75)$$

$$\lambda = \frac{\hbar^2}{2m_0}(A_1 k_z^2 + A_2 k_t^2) + \lambda_\varepsilon \qquad (2.76)$$

$$\theta = \frac{\hbar^2}{2m_0}(A_3 k_z^2 + A_4 k_t^2) + \theta_\varepsilon \qquad (2.77)$$

$$K_t = \frac{\hbar^2}{2m_0} A_5 k_t^2 \qquad (2.78)$$

$$H_t = \frac{\hbar^2}{2m_0} A_6 k_t k_z \qquad (2.79)$$

$$\Delta = \sqrt{2}\Delta_3 \qquad (2.80)$$

($E_v^0 = 0$, $k_t^2 = k_x^2 + k_y^2$, $\varepsilon_{ij} = 0$ for $i \neq j$). The hole effective mass parameters A_1 to A_6 are listed in Table 2.7. Analytical solutions can be derived for the valence band dispersion [31]

$$E_{hh}(k_t, k_z) = S_1 + S_2 - \frac{C_2}{3} \qquad (2.81)$$

$$E_{lh}(k_t, k_z) = -\frac{S_1 + S_2}{2} - \frac{C_2}{3} - i\frac{\sqrt{3}}{2}(S_1 - S_2) \qquad (2.82)$$

$$E_{ch}(k_t, k_z) = -\frac{S_1 + S_2}{2} - \frac{C_2}{3} + i\frac{\sqrt{3}}{2}(S_1 - S_2), \qquad (2.83)$$

where

$$S_1 = \left(R + \sqrt{Q^3 + R^2}\right)^{1/3} \qquad (2.84)$$

$$S_2 = \left(R - \sqrt{Q^3 + R^2}\right)^{1/3} \qquad (2.85)$$

$$R = \frac{1}{6}(C_1 C_2 - 3C_0) - \frac{1}{27}C_2^3 \qquad (2.86)$$

$$Q = \frac{1}{3}C_1 - \frac{1}{9}C_2^2 \qquad (2.87)$$

and

$$C_0 = -\det\left[H^U\right] \qquad (2.88)$$

$$C_1 = FG + G\lambda + F\lambda - \Delta^2 - K_t^2 - 2H_t^2 \qquad (2.89)$$

$$C_2 = -(F + G + \lambda) \qquad (2.90)$$

Figure 2.13 shows the valence band structure of GaN as calculated from the above equations for different strain conditions. At the Γ point, the hole effective

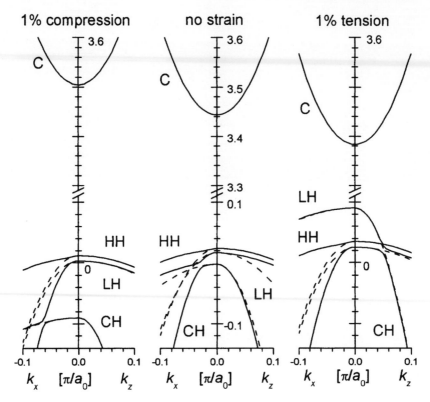

Figure 2.13: GaN energy bands for different strain conditions as calculated from Eqs. (2.64) to (2.90) (C, conduction band; HH, heavy hole band; LH, light hole band; CH, Crystal-field split-hole band). The vertical energy axis is identical in all three cases and given in electron volts. The dashed lines show the parabolic approximation using effective masses at the Γ point (Eqs. (2.91)–(2.96)).

masses can be approximated as [30]

$$m_{hh}^z = -m_0(A_1 + A_3)^{-1} \tag{2.91}$$

$$m_{hh}^t = -m_0(A_2 + A_4)^{-1} \tag{2.92}$$

$$m_{lh}^z = -m_0\left[A_1 + \left(\frac{E_{lh}^0 - \lambda_\varepsilon}{E_{lh}^0 - E_{ch}^0}\right)A_3\right]^{-1} \tag{2.93}$$

$$m_{lh}^t = -m_0\left[A_2 + \left(\frac{E_{lh}^0 - \lambda_\varepsilon}{E_{lh}^0 - E_{ch}^0}\right)A_4\right]^{-1} \tag{2.94}$$

$$m_{ch}^z = -m_0 \left[A_1 + \left(\frac{E_{ch}^0 - \lambda_\varepsilon}{E_{ch}^0 - E_{lh}^0} \right) A_3 \right]^{-1} \qquad (2.95)$$

$$m_{ch}^t = -m_0 \left[A_2 + \left(\frac{E_{ch}^0 - \lambda_\varepsilon}{E_{ch}^0 - E_{lh}^0} \right) A_4 \right]^{-1} . \qquad (2.96)$$

The resulting parabolas are plotted as dashed lines in Fig. 2.13, indicating that some of these effective masses are not valid at larger k values, where the following approximations are more reasonable [30]

$$m_{hh}^z = -m_0 (A_1 + A_3)^{-1} \qquad (2.97)$$

$$m_{hh}^t = -m_0 (A_2 + A_4 - A_5)^{-1} \qquad (2.98)$$

$$m_{lh}^z = -m_0 (A_1 + A_3)^{-1} \qquad (2.99)$$

$$m_{lh}^t = -m_0 (A_2 + A_4 + A_5)^{-1} \qquad (2.100)$$

$$m_{ch}^z = -m_0 A_1^{-1} \qquad (2.101)$$

$$m_{ch}^t = -m_0 A_2^{-1} . \qquad (2.102)$$

2.3 Quantum Wells

A quantum well — or any other variation of the semiconductor composition in growth direction z — can be included in the Schrödinger equation by adding an appropriate potential $V_{qw}(z)$ to the periodic potential $V(\vec{r})$ of the crystal lattice. Within the effective mass approximation, the wave function is given by $\Psi_n(\vec{r}) = \Phi(z) u_{n0}(\vec{r})$ and the envelope function $\Phi(z)$ is the solution to the one-dimensional Schrödinger equation

$$\left[-\frac{\hbar^2}{2} \frac{\delta}{\delta z} \frac{1}{m_n^z(z)} \frac{\delta}{\delta z} + V_{qw}(z) \right] \Phi(z) = E(k_z) \Phi(z), \qquad (2.103)$$

where the periodic crystal potential $V(\vec{r})$ is represented by the effective mass $m_n^z(z)$, which is different for well and barrier materials (n, band index). The procedure to solve Eq. (2.103) is described in several textbooks (see, e.g., [20, 39]). It results in discrete subband energies E_m corresponding to discrete wave numbers $k_{z,m}$ ($m = 1, 2, 3, \ldots$). For a very deep well of thickness d_z the eigenvalues E_m

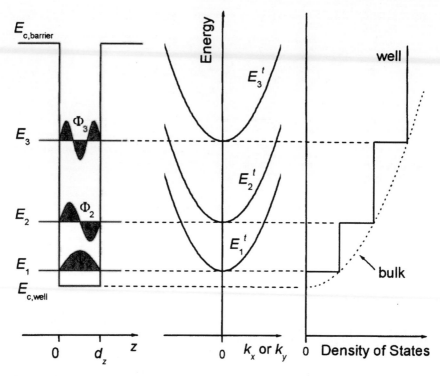

Figure 2.14: Energy levels E_m, wave functions Φ_m, in-plane energy dispersion $E_m^t(k_x, k_y)$, and density of states for a quantum well with infinitively high barrier.

and eigenfunctions Φ_m are

$$E_m = \frac{\hbar^2}{2m_n^z}k_{z,m}^2 = \frac{\hbar^2}{2m_n^z}\left(\frac{m\pi}{d_z}\right)^2 \tag{2.104}$$

$$\Phi_m(z) = \begin{cases} \sqrt{\frac{2}{d_z}}\sin\left(\frac{m\pi}{d_z}z\right) & \text{for even } m \\ \sqrt{\frac{2}{d_z}}\cos\left(\frac{m\pi}{d_z}z\right) & \text{for odd } m \end{cases} \tag{2.105}$$

(for $V_{\text{qw}}(z) = 0$ inside and $V_{\text{qw}}(z) \to \infty$ outside the well). These solutions are plotted in Fig. 2.14. While only discrete wave numbers $k_{z,m}$ are allowed in the z direction, the in-plane quantum well energy dispersion is parabolic near the Γ point (Fig. 2.14):

$$E_m^t(k_x, k_y) = E_m + \frac{\hbar^2}{2m_{n,m}^t}(k_x^2 + k_y^2). \tag{2.106}$$

In effective mass approximation, the two-dimensional density of electron states within each subband m equals

$$D_m(E) = \frac{m^t_{n,m}}{\pi \hbar^2 d_z} \quad \text{for } E > E_m, \tag{2.107}$$

resulting in a step-like increase with rising energy as higher subbands overlap (Fig. 2.14).

Typical semiconductor quantum wells are rather shallow, which causes the wave function to penetrate into the barrier region. The distance of the energy levels E_m shrinks with lower barriers; however, the dependence $E_m = m^2 E_1$ is maintained. The actual potential depths ΔE_c and ΔE_v of semiconductor quantum wells are critical parameters in device simulations but they often are not exactly known. Section 2.5 outlines several methods for estimating the band offsets.

The effective mass approximation given in Eq. (2.106) is reasonably accurate only for decoupled bands like the conduction band. For the coupled valence bands, the three-dimensional Schrödinger equation needs to be solved numerically, subtracting the quantum well potential $V_{qw}(z)$ from the diagonal elements of the Hamiltonian matrix in Eq. (2.36). Several numerical methods, for instance, the propagation matrix method [21], the finite element method [40], and the finite difference method [41] can be employed. Figure 2.15 illustrates the in-plane band structure for different GaAs-based quantum wells with $Al_x Ga_{1-x} As$ barriers. Alloy materials like AlGaAs require the interpolation of binary material parameters as described in Section 2.4. An unstrained 8 nm wide GaAs quantum well with $Al_{0.15}Ga_{0.85}As$ barriers is evaluated in Fig. 2.15a. The same quantum well is used for the device example in Chap. 11. Energy quantum levels are at different positions for heavy and light holes due to the different masses m^z_{hh} and m^z_{lh}. Valence mixing effects lead to strongly nonparabolic dispersion functions with sometimes negative effective masses at the Γ point (LH1, HH3). Figure 2.15b shows the same quantum well at reduced thickness, which moves the energy levels toward the barrier edge, reducing their total number (cf. Eq. (2.104)). In fact, the second subband is barely confined for heavy holes (HH2). Dashed lines show the parabolic dispersion obtained from the bulk masses m^t_{hh} and m^t_{lh}, which gives poor agreement. Case (c) describes a 8-nm-wide GaAs quantum well with $Al_{0.3}Ga_{0.7}As$ barrier, which allows for the same number of energy levels as in (a). However, the higher barrier slightly moves all levels further away from the well edge at 0 eV. Finally, Fig. 2.15d shows a strained $In_{0.05}Ga_{0.95}As$ quantum well where the well edge is different for heavy and light holes (cf. Eqs. (2.40) and (2.41)). Therefore, only one light-hole subband is supported. The very small barrier strain is neglected in all four examples. These quantum wells shall be further investigated in Chap. 5.

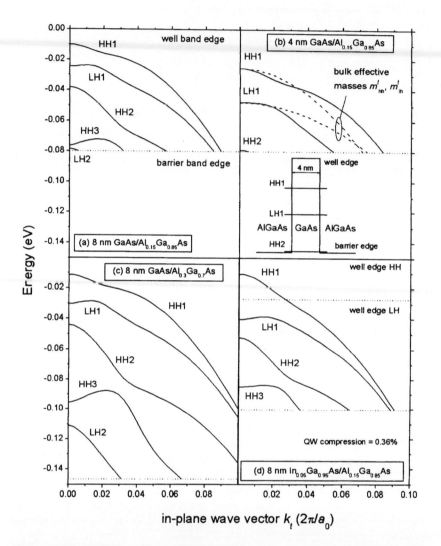

Figure 2.15: In-plane energy dispersion $E(k_t)$ for valence subbands in four slightly different quantum wells, grown on GaAs with AlGaAs barriers (HH#, heavy hole; LH#, light hole; #, subband number). The edge of the quantum well is at $E = 0$. The dotted line marks the barrier band edge, beyond which the energy states are extended. Effective mass approximations using Eqs. (2.48) and (2.49) are indicated by dashed lines in (b). The GaAs wells are unstrained, compressive strain in (d) splits the well edges for heavy and light holes.

2.4 Semiconductor Alloys

Optoelectronic devices often employ alloys of binary materials. Because of the random distribution of elements from the same group within the alloy lattice, exact calculations of material parameters are hardly possible. Instead, material parameters of quaternary alloys (Q) are predicted by interpolation of known binary (B) or ternary (T) alloy data. Parameters of ternary alloys AB_xC_{1-x} are usually given as

$$T_{ABC}(x) = xB_{AB} + (1-x)B_{AC} - x(1-x)C_{ABC} \qquad (2.108)$$

using an empirical bowing parameter C_{ABC}. According to Vegard's law, $C_{ABC} = 0$ for the lattice constant. Bowing is also not pronounced for the effective mass of most alloys. However, strong bowing is often observed for the energy gap. Experimentally determined bowing parameters for the direct band gap and the spin–orbit splitting are listed in Table 2.8. It should be noted that the smallest band gap may change from direct to indirect as composition is varied. This is the case for $Al_xGa_{1-x}As$ at $x = 0.42$ (Fig. 2.16). Furthermore, strain causes an additional change of the band gap energy (see Section 2.2.2).

Based on empirical interpolation formulas for ternary compounds, properties of quaternary alloys $A_xB_{1-x}C_yD_{1-y}$ can be interpolated as

$$Q(x, y) = [x(1-x)[yT_{ABC}(x) + (1-y)T_{ABD}(x)]$$
$$+ y(1-y)[xT_{ACD}(y) + (1-x)T_{BCD}(y)]]/[x(1-x) + y(1-y)]. \qquad (2.109)$$

Alternatively, alloys of the type $AB_xC_yD_{1-x-y}$ are described by

$$Q(x, y) = \frac{xyT_{ABC}(u) + y(1-x-y)T_{ACD}(v) + x(1-x-y)T_{ABD}(w)}{xy + y(1-x-y) + x(1-x-y)} \qquad (2.110)$$

with

$$u = \frac{1-x+y}{2} \qquad v = \frac{2-x-2y}{2} \qquad w = \frac{2-2x-y}{2}. \qquad (2.111)$$

As an example, Fig. 2.17 shows the direct band gap versus composition for various quaternary III–V semiconductors lattice-matched to InP.

2.5 Band Offset at Heterointerfaces

The model-solid theory is often employed to estimate the band edge offsets ΔE_c and ΔE_v of cubic semiconductors [23]. It lines up the band structure of different semiconductors by introducing an average valence band energy $E^0_{v,av}$ for

Table 2.8: Bowing Parameter in Eq. (2.108) for Energy Gaps E_g^Γ, E_g^X, E_g^L, Valence Band Edge E_v^0, and Spin–Orbit Splitting Δ_0 at Room Temperature [13, 42, 43]

Parameter Unit	$C(E_g^\Gamma)$ (eV)	$C(E_g^X)$ (eV)	$C(E_g^L)$ (eV)	$C(E_v^0)$ (eV)	$C(\Delta_0)$ (eV)
Al(As,Sb)	0.84	0.28	0.28	−1.71	0.15
Al(P,Sb)	3.56	2.7	2.7		
Al(P,As)	0.22	0.22	0.22		0.0
Ga(As,Sb)	1.43	1.2	1.2	−1.06	0.6
Ga(P,Sb)	2.558	2.7	2.7		
Ga(P,As)	0.19	0.24	0.16		0.0
In(As,Sb)	0.67	0.6	0.6		1.2
In(P,Sb)	1.9	1.9	1.9		0.75
In(P,As)	0.10	0.27	0.27		0.16
(Al,Ga)P	0.0	0.13			0.0
(Al,In)P	−0.48	0.38			−0.19
(Ga,In)P	0.65	0.2	1.03		0.0
(Al,Ga)As	−0.127+1.31x	0.055	0.0		0.0
(Al,In)As	0.70	0.0		−0.64	0.15
(Ga,In)As	0.477	1.4	0.33	−0.38	0.15
(Al,Ga)Sb	−0.044+1.22x	0.0	0.0		0.3
(Al,In)Sb	0.43				0.25
(Ga,In)Sb	0.415	0.33	0.4		0.1

Note. Wurtzite nitride bowing is discussed in Chap. 9.

unstrained semiconductors, which is $\Delta_0/3$ below the valence band edge:

$$E_{v,av}^0 = E_v^0 - \frac{\Delta_0}{3} \qquad (2.112)$$

(Fig. 2.18). The numbers $E_{v,av}^0$ are listed in Table 2.9, and they are obtained by calculating the average electrostatic potential for model-solids of neutral atoms. The position of the conduction band edge is calculated by

$$E_c^0 = E_v^0 + E_g = E_{v,av}^0 + \frac{\Delta_0}{3} + E_g. \qquad (2.113)$$

For alloys $A_x B_{1-x} C$, the interpolation between the binary values $E_{v,av}^0$ should consider that the alloy lattice constant a_{ABC} is different from the binary lattice

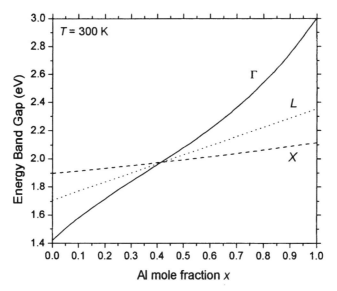

Figure 2.16: Direct (Γ) and indirect (L, X) band gaps of $Al_xGa_{1-x}As$ as a function of composition x at room temperature [13].

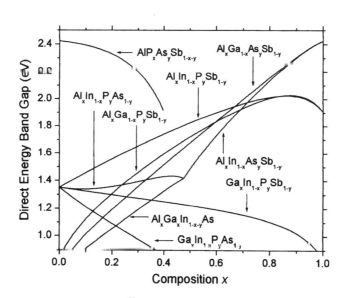

Figure 2.17: Direct band gap E_g^{Γ} bowing for quaternary III–V compounds lattice-matched to InP.

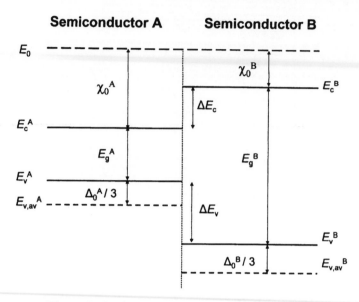

Figure 2.18: Illustration of band offset parameters.

constants a_{AC} and a_{BC}. This deformation of the binary model-solid is accounted for by the bowing parameter (see Eq. (2.108)) [44]

$$C_{ABC}(E^0_{v,av}) = 3[a_v(AC) - a_v(BC)]\frac{a_{AC} - a_{BC}}{a_{ABC}} \qquad (2.114)$$

using binary deformation potentials a_v as listed in Table 2.6. This bowing parameter itself is a function of alloy composition. Under strain, the average energy $E_{v,av}$ is shifted from its unstrained value $E^0_{v,av}$ as

$$E_{v,av} = E^0_{v,av} - P_\varepsilon = E^0_v - \frac{\Delta_0}{3} - P_\varepsilon. \qquad (2.115)$$

By comparison with fully self-consistent interface calculations, the error bar on band offsets determined from the model-solid theory extends up to ± 0.1 eV [45].

Many other theoretical and experimental investigations of heterostructure band offsets have been published, leading to a list of typical values for the valence band edge E^0_v in unstrained bulk material (Table 2.9) [13]. The absolute energy scale is arbitrary, here the InSb valence band edge is used as origin. Known bowing parameters are listed in Table 2.8. Strain effects are implemented by using the equations given in Sections 2.2.2 and 2.2.4.

Alternatively, the lowest conduction band edge E^0_c of unstrained semiconductors is sometimes used as a reference level by employing the electron affinity χ_0,

Table 2.9: Valence Band Edge Reference Level E_v^0 [13], Split-Off Energy Δ_0 [13, 23], Average Valence Band Energy $E_{v,av}^0$ [23], and Electron Affinity χ_0 [46] for Unstrained Cubic Semiconductors

Parameter Unit	E_v^0 (eV)	Δ_0 (eV)	$E_{v,av}^0$ (eV)	χ_0 (eV)
Si		0.045	−7.03	4.01 (X)
Ge		0.297	−6.35	4.13 (L)
GaAs	−0.8	0.341	−6.92	4.07 (Γ)
InP	−0.49	0.108	−7.04	4.40 (Γ)
AlAs	−1.33	0.28	−7.49	3.85 (X)
GaSb	−0.03	0.76	−6.25	4.06 (Γ)
AlSb	−0.41	0.676	−6.66	3.60 (X)
InAs	−0.59	0.39	−6.67	4.90 (Γ)
GaP	−1.27	0.08	−7.40	4.0 (X)
AlP	−1.74	0.07	−8.09	3.98 (X)
InSb	0.0	0.81	−6.09	4.59 (Γ)
ZnS		0.07	−9.15	3.90 (Γ)
ZnSe		0.43	−8.37	4.09 (Γ)
ZnTe		0.91	−7.17	3.53 (Γ)
CdS		0.064		4 87 (Γ)
CdSe		0.416		4.95 (Γ)
CdTe		0.93	−7,07	4 78 (Γ)

Note. The origin of the energy axis varies between E_v^0, $E_{v,av}^0$, and χ_0 as differences within each column are of interest only. Numbers for nitride materials are listed in Table 2.7.

which defines the energy difference to the universal vacuum energy E_0:

$$E_c^0 = E_0 - \chi_0. \qquad (2.116)$$

Affinity values are listed in Table 2.9. Strain effects on the conduction band edge at the Γ point are described by Eq. (2.42). Bowing parameters for the affinity can be calculated by weighting the band gap bowing parameter $C(E_g)$ with the average conduction band offset ratio $\Delta E_c / \Delta E_g$ of the binary endpoint materials for the same symmetry point:

$$C(\chi_0) = -C(E_g)\frac{\Delta E_c}{\Delta E_g}. \qquad (2.117)$$

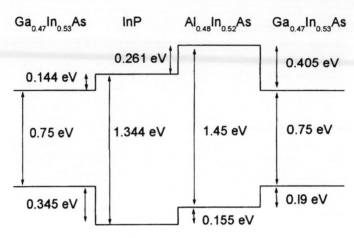

Figure 2.19: Band offsets for $Ga_{0.47}In_{0.53}As$ and $Al_{0.48}In_{0.52}As$ having the same lattice constant as InP ($T = 300\,K$).

Typical values used for the conduction band offset ratio $\Delta E_c / \Delta E_g$ at the Γ point are

0.65	for $Al_x Ga_{1-x}As/GaAs$
0.4	for $Ga_x In_{1-x}As_y P_{1-y}/InP$
0.7	for $Al_x Ga_y In_{1-x-y}As/InP$
0.7	for $Al_x Ga_y In_{1-x-y}N/GaN$.

However, such average numbers only give rough approximations, and research publications should be consulted for specific alloy interfaces (see review in [13]). Offsets for common alloys, which are lattice-matched to InP, are shown in Fig. 2.19.

Further Reading

- S. L. Chuang, *Physics of Optoelectronic Devices*, Wiley, New York, 1995.
- K. W. Böer, *Survey of Semiconductor Physics*, Vol. I, Wiley, New York, 2002.
- J. P. Loehr, *Physics of Strained Quantum Well Lasers*, Kluwer, Boston, 1998.
- R. Enderlein and N. J. M. Horing, *Fundamentals of Semiconductor Physics and Devices*, World Scientific, Singapore, 1997.
- N. W. Ashcroft and N. D. Mermin, *Solid State Physics*, Saunders, Forth Worth, 1976.

Chapter 3

Carrier Transport

Electrical current flow in semiconductors is mainly dominated by drift and diffusion of electrons and holes. This chapter reviews the physical mechanisms that affect the carrier transport in optoelectronic devices, including pn-junctions, thermionic emission at heterojunctions, and tunneling. Simple formulas for the carrier mobility and for various carrier generation and recombination mechanisms are given. Advanced transport models are outlined at the end.

3.1 Drift and Diffusion

Semiconductor device simulation software most commonly uses the drift–diffusion model to compute the flow of electrons and holes. Drift current is generated by an electric field $\vec{F}$ and it is proportional to the conductivity of electrons $\sigma_n = q\mu_n n$ and holes $\sigma_p = q\mu_p p$. Diffusion current is driven by the concentration gradient of electrons ∇n and holes ∇p. It is proportional to the diffusion coefficient D_n and D_p, respectively. For uniform semiconductors, the total current density of electrons and holes is written as

$$\vec{j}_n = q\mu_n n\vec{F} + qD_n\nabla n \tag{3.1}$$

$$\vec{j}_p = q\mu_p p\vec{F} - qD_p\nabla p. \tag{3.2}$$

We consider the elementary charge q to always be a positive number, so that the current flows in the direction of the electric field while the electrons move in opposite direction. Both carriers diffuse "downhill" toward lower carrier concentration while ∇n and ∇p point in the "uphill" direction.

Changes in the local carrier concentration in time must be accompanied by a spatial change in current flow $(\nabla \vec{j})$ and/or by the generation (rate G) or recombination (rate R) of electron–hole pairs. This relation is expressed by the continuity equations

$$q\frac{\partial n}{\partial t} = \nabla \cdot \vec{j}_n - q(R - G) \tag{3.3}$$

$$q\frac{\partial p}{\partial t} = -\nabla \cdot \vec{j}_p - q(R - G). \tag{3.4}$$

49

Finally, the electric field itself is affected by the charge distribution, which includes mobile (n, p) and fixed charges (dopant ions p_D, n_A). This relationship is described by the Poisson equation

$$\nabla \cdot (\varepsilon_0 \varepsilon_{st} \vec{F}) = q(p - n + p_D - n_A). \tag{3.5}$$

The net charge inside the device is zero; i.e., mobile and fixed charges compensate each other.

This set of equations constitutes the drift–diffusion model for the electrical behavior of semiconductor devices; more general models are discussed at the end of this chapter. The equations deliver the unknown functions $n(\vec{r}, t)$, $p(\vec{r}, t)$, and $\vec{F}(\vec{r}, t)$ if appropriate initial and boundary conditions are considered (see Section 3.5). Inside a semiconductor device, these quantities can vary by many orders of magnitude. They are sometimes replaced by the electrostatic potential ϕ and the quasi-Fermi levels E_{Fn} and E_{Fp}, which shall be introduced below and which exhibit less dramatic variations.

3.2 pn-Junctions

At the junction of n-doped and p-doped semiconductor regions, both carriers diffuse into the other region and recombine, creating a transition region that is depleted of carriers (depletion width $= w_n + w_p$, see Fig. 3.1). The remaining ionized donors and acceptors build an electric field that counteracts the carrier diffusion. With balanced drift and diffusion, no net current is flowing. According to the Poisson equation, the slope of the electric field $\nabla \vec{F}$ is proportional to the net charge; hence the maximum field occurs at the pn-junction. Electrostatic field $\vec{F}$ and potential ϕ are related by

$$\vec{F} = -\nabla \phi; \tag{3.6}$$

i.e., the slope of the potential is given by the electric field. In other words, the potential does not change in regions with zero field (and the field does not change in regions without charges — see Fig. 3.1). The built-in field relates to a built-in potential ϕ_{bi} and to a change of the band edge energies $\Delta E = -q \Delta \phi_{bi}$ across the junction.

Without external bias, the Fermi level E_F is a flat line. The Fermi function represents the electron distribution in thermal equilibrium, with zero net transfer of electrons between different energy levels. The equilibrium can be disturbed by external forces that provide additional electrons and/or holes. Providing electrons from the right and holes from the left in Fig. 3.1 (forward bias) generates excess carriers in the depletion region that cannot immediately find a recombination partner. This nonequilibrium carrier distribution in valence and conduction band is not described by the same Fermi function anymore. Assuming equilibrium within

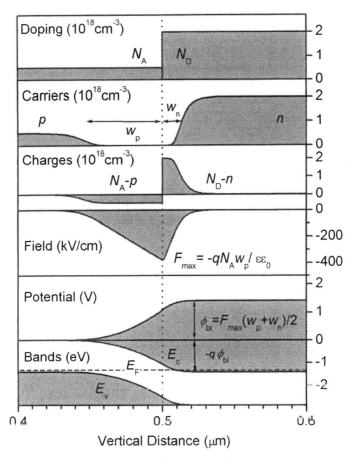

Figure 3.1: Physical parameter profiles of an abrupt GaAs pn-junction (ϕ_{bi}, built-in potential).

each band, the excess carriers lead to the separation of Fermi levels for electrons and holes. These quasi-Fermi levels E_{Fn} and E_{Fp} are illustrated in Fig. 3.2 for the same pn-junction (note the larger length scale). Excess electrons and excess holes penetrate some distance into the other region until recombination establishes an equilibrium with single Fermi level. The diffusion length is proportional to the square root of the carrier lifetime before recombination (see Section 3.7).

3.3 Heterojunctions

Most modern optoelectronic devices utilize heterojunctions between different semiconductor materials. They are used to restrict the flow of carriers or to collect

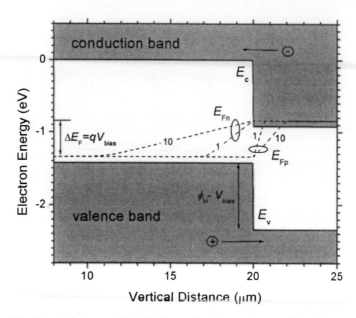

Figure 3.2: Band diagram of the GaAs *pn*-junction at forward bias $V_{bias} = 0.5$ V. Dashed lines give the quasi-Fermi levels with the carrier lifetime (ps) as the parameter (E_{Fp}, holes; E_{Fn}, electrons). Extremely short lifetimes are used here for illustration; minority lifetimes up to $\tau_n = 5$ ns and $\tau_p = 3$ μs have been measured in GaAs [47].

carriers where they are needed. The band diagram of an *n*-doped AlGaAs/GaAs heterojunction is plotted in Fig. 3.3. The most important junction parameters are the offsets ΔE_c and ΔE_v of conduction and valence band edge, respectively. They add up to the band gap difference ΔE_g between both materials. The band offset varies with the material system and it is often not exactly known. More details on the band offset are given in Section 2.5.

   The band bending near the heterojunction is caused by the redistribution of electrons toward the material with the lower electron energy. The remaining donor ions pull the electrons back and an equilibrium carrier profile is established. In Fig. 3.3, the AlGaAs side is depleted of electrons, which translates into a band edge E_c that is well above the Fermi level E_F. Electrons accumulate on the low-band-gap GaAs side as shown by the penetration of the Fermi level into the conduction band. Without net current flow, the Fermi level is perfectly flat. It is the same on both sides of the interface; i.e., the interface Fermi distribution of electrons is identical on both sides. In other words, the same number of electrons penetrates the interface from the left- and right-hand sides, resulting in zero net current. To achieve net

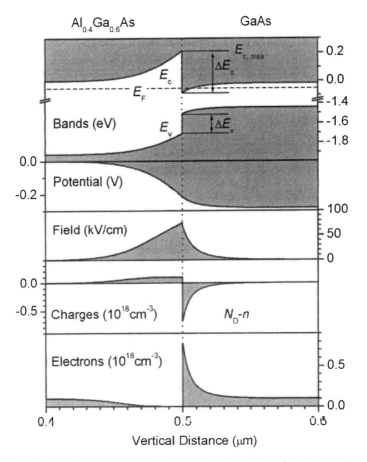

Figure 3.3: Vertical parameter profiles at an $Al_{0.4}Ga_{0.6}As/GaAs$ heterointerface with uniform n-doping of 10^{17} cm^{-3}.

current across the junction, the carrier flow in both directions must be asymmetric; i.e., the electron quasi-Fermi levels on both sides of the interface must be different ($E_{Fn,left} \neq E_{Fn,right}$). This is expressed by the Richardson formula for thermionic electron emission across a barrier (similar for holes):

$$j_n = A_n^* T^2 \left\{ \exp\left[\frac{E_{Fn,left} - E_{c,max}}{k_B T} \right] - \exp\left[\frac{E_{Fn,right} - E_{c,max}}{k_B T} \right] \right\} \qquad (3.7)$$

with the Richardson constant

$$A_n^* = \frac{q m_c k_B^2}{2\pi^2 \hbar^3} \qquad (3.8)$$

($m_{\rm c}$, average effective mass; $A_n^* = 120\,{\rm A\,cm^{-2}\,K^{-2}}$ for $m_{\rm c} = m_0$). The exponential factors describe the electron concentration on both sides in Boltzmann approximation, which is valid for $E_{\rm c,max} - E_{\rm F} > 3k_{\rm B}T$. The electron concentrations are identical and the net current vanishes for $E_{Fn,{\rm left}} = E_{Fn,{\rm right}}$.

The current flow across a heterojunction can be limited by thermionic emission or by drift and diffusion. The latter case arises when the junction depletion width is much larger than the carrier "free flight" distance between two scattering events (see Section 3.6). In that case, carrier penetration is mainly hindered by the high resistance of the depleted region, and inclusion of the thermionic emission formula in the drift–diffusion equations is not important. In the opposite case of small depletion width, the electrons hit the barrier in "free flight" and Eq. (3.7) needs to be included. With very thin depletion region (high doping), electron tunneling through the potential barrier may create an even stronger current than thermionic emission.

3.4 Tunneling

Tunneling is a quantum mechanical phenomenon that results from the wave nature of electrons. When an electron wave arrives at an energy barrier, one part of it gets reflected and another part penetrates the barrier. The penetration depth depends on the barrier height. If the barrier is thin enough, part of the electron wave continues on the other side (Fig. 3.4). The tunnel probability or transparency $T(E)$ of the

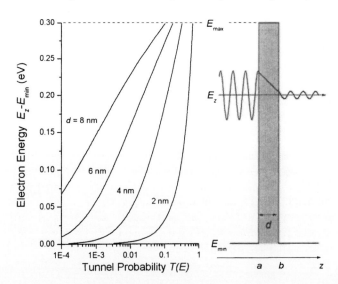

Figure 3.4: Tunneling of an electron wave in GaAs through a rectangular barrier: tunnel transparency $T(E)$ with barrier width $d = b - a$.

barrier strongly depends on the electron energy E and on the barrier shape $E_{bar}(z)$. The transparency $T(E)$ is the same in both directions; i.e., a net tunnel current arises only with different carrier distributions on both sides. The total tunnel current is the sum of the tunnel currents at all allowed energy levels below the barrier peak E_{max}. In calculating the tunnel current, all three dimensions of electron motion need to be considered on both sides. For tunneling between two bulk regions with parabolic energy bands, the tunnel current density can be written as

$$j_{tun} = \frac{q m_n k_B T}{2\pi^2 \hbar^3} \int_{E_{min}}^{E_{max}} dE_z \, T(E_z) \ln \left\{ \frac{1 + \exp[-(E_z - E_{Fn,left})/k_B T]}{1 + \exp[-(E_z - E_{Fn,right})/k_B T]} \right\}, \tag{3.9}$$

where E_z is the energy of motion perpendicular to the tunnel junction (similar for holes). This equation neglects the difference between the effective masses on both sides of the barrier [48]. Note that the tunnel current vanishes for $E_{Fn,left} = E_{Fn,right}$.

For arbitrary barriers $E_{bar}(z)$, the tunnel transparency needs to be calculated numerically by solving the Schrödinger equation. Analytical solutions only exist for idealized simple barrier shapes, e.g., for a rectangular barrier as shown in Fig. 3.4. With a maximum barrier energy E_{max} and thickness d of the rectangular barrier, the tunneling transparency is given by

$$T(E_z) = \left[1 + \frac{\sinh^2\left(d\sqrt{\frac{2m_u}{\hbar}(E_{max} - E_z)} \right)}{4\frac{E_z - E_{min}}{E_{max} - E_{min}}\left(1 - \frac{E_z - E_{min}}{E_{max} - E_{min}} \right)} \right]^{-1}. \tag{3.10}$$

Results are plotted in Fig. 3.4 for several barrier widths d. The tunnel probability is high only for very thin barriers. Similar barriers occur in real optoelectronic devices, for instance, between quantum wells. However, the internal electrical field changes the barrier shape. Furthermore, only discrete electron energy levels exist within neighboring quantum wells, which need to line up for tunneling to occur (resonant tunneling [49]). Thus, Eq. (3.10) only serves as an illustration here; accurate solutions for real devices require numerical techniques.

The propagation matrix method is often used to account for tunneling through arbitrary heterobarriers in semiconductor devices [50]. This method slices up the barrier into many layers with piecewise constant potential; i.e., it replaces the real shape $E_{bar}(z)$ by a multiple-step function. The smaller the steps, the higher the accuracy but the longer the computation time.

A slowly varying barrier profile $E_{bar}(z)$ allows for the WKB approximation of the tunnel transparency

$$T(E_z) = \exp\left\{-\frac{2}{\hbar}\int_a^b dz\sqrt{2m_n[E_{bar}(z) - E_z]}\right\}, \qquad (3.11)$$

where a and b, respectively, specify the location at which the electron enters and leaves the barrier ($E_{bar}(a) = E_z$, $E_{bar}(b) = E_z$). However, WKB approximations should be applied with great care. If the requirement of a slowly varying potential is not satisfied, like at heterojunctions, significant error may arise [51].

Tunneling is a very rich and diverse phenomenon, especially when it comes to real device applications. Examples are heterostructure effects on the wave function [52] and phonon-assisted (inelastic) tunneling [53]. More details can be found in specialized texts, e.g., in [54].

3.5 Boundary Conditions

At the device boundaries, the semiconductor is in contact with metals or with insulating materials. For both cases, boundary conditions are needed for the three unknown functions in the Poisson equation ($\vec{F}$ or ϕ) and continuity equations (n, p or E_{Fn}, E_{Fp} or j_n, j_p). Appropriate boundary conditions are essential for successful device simulations. Boundary conditions also represent the external bias applied to the device. In general, three types of boundary conditions can be distinguished:

- *Dirichlet* boundary conditions give explicit values for the unknown functions at the boundary.
- *Neumann* boundary conditions give explicit values for the derivative (slope) of the unknown functions perpendicular to the boundary, which may correspond to the flux of particles or energy across the boundary.
- *Mixed* boundary conditions contain combinations of absolute and slope values.

In addition to real physical boundaries, device simulations often use artificial boundaries to reduce the size of the device region simulated. Such boundaries may represent symmetry planes of the device not penetrated by the flow of carriers or energy. In this case, zero-slope Neumann conditions are appropriate. Otherwise, artificial boundaries should be far away from the active parts of the device in order to not affect the simulated device performance.

In the following, simple examples of steady-state boundary conditions for three common types of physical interfaces are described. The direction normal to the boundary is given by the vector $\vec{\nu}$, which may be identical to the x, y, or z direction.

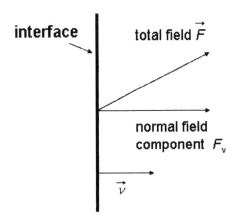

Figure 3.5: Illustration of the normal vector component at interfaces.

The normal component of the electrostatic field, for instance, is given by $F_\nu = \vec{F}\vec{v}$ (Fig. 3.5). It is here assumed to vanish on the outer side of the boundary.

3.5.1 Insulator–Semiconductor Interface

Interfaces between semiconductors and dielectric materials (including air and vacuum) exhibit a high density of energy levels within the band gap. These levels are created by the interruption of the crystal periodicity as well as by other defects. Interface defects may trap carriers and build a significant boundary charge density Q_b. These charges define the normal electrostatic field at the boundary by the Dirichlet condition

$$-\varepsilon_0\varepsilon_{st}\vec{F}\vec{v} = Q_b;\qquad(3.12)$$

the field vanishes without interface charges. This equation is identical to a Neumann condition for the electrostatic potential ϕ:

$$\varepsilon_0\varepsilon_{st}\frac{\partial\phi}{\partial\vec{v}} = Q_b.\qquad(3.13)$$

Interface defects may also enhance electron–hole recombination, characterized by a surface recombination velocity v_{sr}, which draws the interface recombination current density

$$\vec{j}_p\vec{v} = -\vec{j}_n\vec{v} = qv_{sr}(np - n_0p_0).\qquad(3.14)$$

The surface recombination velocity is a material parameter that depends on the surface treatment (e.g., passivation). Typical values for nonpassivated surfaces are

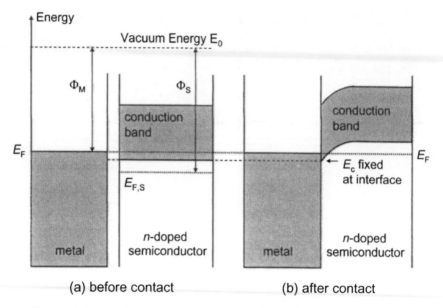

Figure 3.6: Ohmic contact between metal and an n-type semiconductor.

on the order of 10^5 cm/s for GaAs and 10^4 cm/s for InP. However, boundary defects can often be ignored if these boundaries are far away from the active part of the device.

3.5.2   *Metal–Semiconductor Contact*

For metals and semiconductors, the work function $\Phi = E_0 - E_F$ is the average energy required to remove an electron from the crystal (E_0, vacuum energy, see Fig. 3.6). Work functions for various metals are listed in Table 3.1. The work function of semiconductors depends on the doping (cf. Fig. 1.6). The metal–semiconductor work function difference

$$\Phi_{MS} = \Phi_M - \Phi_S \qquad (3.15)$$

can be negative or positive, which, together with the doping type, determines the nature of the contact: Ohmic contact with linear current–voltage relation or Schottky contact with current rectifying properties like a diode.

3.5.2.1   Ohmic Contact

Figure 3.6 shows an n-type semiconductor brought into contact with a metal of lower work function ($\Phi_{MS} < 0$). Some metal electrons consequently lower their

Table 3.1: Work functions Φ_M of Selected Metals in Electron Volts (eV) [55]

Metal	Φ_M	Metal	Φ_M	Metal	Φ_M	Metal	Φ_M	Metal	Φ_M
Ag (100)	4.75	Fe β	4.62	Mg	3.68	Pb	4.14	Ti	4.45
Ag (111)	4.81	Fe γ	4.68	Mo	4.3	Pt	5.32	W (001)	4.52
Al	4.08	Hg	4.53	Mn	3.76	Sn β	4.50	W (111)	4.39
Au	4.82	In	3.8	Ni	5.01	Snγ	4.38	Zn	4.26

energy by moving into the semiconductor, forming a low-resistance accumulation region.[1] This allows for a simple Dirichlet boundary condition

$$\phi = V_{\text{bias}} - \phi_{\text{bi}} \qquad (3.16)$$

with the applied external bias V_{bias} and the built-in potential ϕ_{bi} (cf. Section 3.2). In fact, the absolute value of the boundary potential is not important, as long as the difference between the contacts is equal to the external bias. A time-dependent bias $V_{\text{bias}}(t)$ requires the time-dependent solution of the semiconductor equations. Equation (3.16) describes voltage-controlled contacts. Alternatively, the contact can be controlled by the applied current I_{con}. In that case, integration over the contact area gives

$$I_{\text{con}} = \int dA(\vec{j}_n + \vec{j}_p)\vec{v}, \qquad (3.17)$$

which can be transformed into a more complicated boundary condition for the potential using Eqs. (3.1) and (3.2). The boundary conditions for the carrier concentrations are obtained by assuming thermal equilibrium and charge neutrality. This allows for the following two conditions to be established at the boundary

$$np - n_0 p_0 = 0 \qquad (3.18)$$

$$p - n + p_D - n_A = 0 \qquad (3.19)$$

with the equilibrium concentrations n_0 and p_0. These two equations can be transformed into explicit Dirichlet conditions for n and p [56].

3.5.2.2 Schottky Contacts

Figure 3.7 shows an n-type semiconductor brought into contact with a metal of higher work function. Due to the positive work function difference Φ_{MS}, electrons

---

[1]A similar situation with hole accumulation arises for p-type semiconductors and $\Phi_{MS} > 0$.

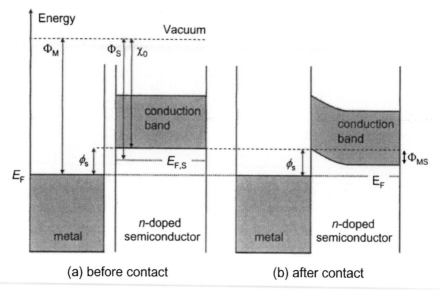

Figure 3.7: Schottky contact between metal and n-doped semiconductor.

are able to lower their energy by moving from the semiconductor into the metal, forming a depletion region.[2] A Schottky barrier of height

$$\phi_s = \Phi_M - \chi_0 \qquad (3.20)$$

is formed, which makes it difficult to inject electrons into the semiconductor (χ_0, electron affinity, see Table 2.9). In opposite direction, the semiconductor surface potential is Φ_{MS} at zero bias and it changes with the applied bias. Thus, the resistance of the Schottky contact depends on the direction of the current flow.

Schottky contacts are difficult to describe mathematically as they involve complicated transport mechanisms like thermionic emission and quantum tunneling (see [57] for a detailed discussion). However, in case they are not essential for the device performance, Schottky contacts are often treated in a strongly simplified way. The surface potential can be given by

$$\phi = V_{\text{bias}} - \phi_s. \qquad (3.21)$$

The carrier concentrations at the contact depend on the current densities. Various current boundary condition can be found in the literature [56, 58] with the

---

[2]A similar situation arises with p-doped semiconductors and $\Phi_{MS} < 0$.

following form often used in device simulation [59]

$$\vec{j}_n \vec{v} = q v_n^{\text{th}} (n - n_0),$$ (3.22)

$$\vec{j}_p \vec{v} = -q v_p^{\text{th}} (p - p_0),$$ (3.23)

where v_n^{th} and v_p^{th} are the thermionic emission velocities

$$v_n^{\text{th}} = \frac{A_n^* T^2}{q N_c}$$ (3.24)

$$v_p^{\text{th}} = \frac{A_p^* T^2}{q N_v}$$ (3.25)

with the effective Richardson constant A^* (cf. Section 3.3). Thermal equilibrium with $n = n_0$ and $p = p_0$ is established by infinite thermionic emission velocities ($E_{Fn} = E_{Fp}$).

3.6 Carrier Mobility

The carrier mobility is the key material parameter in transport simulations. It is limited by collisions of electrons and holes with other carriers, with crystal defects, and with phonons (lattice vibrations). Those scattering events slow down the carriers and constitute the electrical resistance of the material. Using the average carrier "flight" time τ_{sc} between two scatter events, the Drude model simply writes the carrier mobility as

$$\mu = \tau_{\text{sc}} \frac{q}{m_n}$$ (3.26)

with the effective mass m_n. Note that, when combining bands or directions, the mobility effective mass differs from the density-of-states effective mass [2, 7, 9]. The time τ_{sc} depends significantly on the microstructure of the semiconductor, on the electric field, and on the temperature. Theoretical approaches often calculate the scattering time τ_{sc} for specific scatter mechanisms (see reviews in [2, 53]). The total mobility can then be obtained by adding up the results for all relevant scatter mechanisms. This scatter theory approach results in complicated equations that include some physical parameters that are hardly known for many semiconductors. Instead, for practical device simulation, empirical mobility formulas that are fitted to measurements are often used. We shall summarize some common mobility functions in the following. Compared to the vast amount of research on silicon (see reviews in [53, 56]), relatively few mobility formulas have been published for

Table 3.2: Mobility Model Parameters of Eqs. (3.27) and (3.28) at Room Temperature

Symbol Unit	μ_o (cm²/Vs)	μ_{dop} (cm²/Vs)	N_{ref} (10¹⁷cm⁻³)	α —	δ_0 —	δ_{dop} —	δ_1 —	δ_{ref} —	δ_α —
				Electrons					
Si[a]	1430	80	1.12	0.72	-2.0	-0.45		3.2	0.065
Ge[b]	3895	641	0.613	1.04	-1.67	-0.57	-1.66	2.4	-0.146
GaAs[c]	9400	500	0.6	0.394	-2.1			-3.0	
InP[c]	5200	400	3.0	0.42	-2.0			-3.25	
GaP[c]	152	10	44.0	0.80	-1.6			-0.71	
AlP[f]	60								
AlAs[c]	400	10	5.46	1.0	-2.1			-3.0	
GaSb[e]	13000	100	1.0	0.5	-2.1				
AlSb[e,f]	200	50	1.0	0.5	-1.8				
InSb[f]	70000				-1.66				
InAs[c]	34000	1000	11.0	0.32	-1.57			-3.0	
In$_{0.53}$Ga$_{0.47}$As[c]	14000	300	1.3	0.48	-1.59			-3.68	
In$_{0.49}$Ga$_{0.51}$P[c]	4300	400	0.2	0.7	-1.66				
In$_{0.52}$Al$_{0.48}$As[c]	4800	800	0.3	1.1					
GaN[d]	1461	295	1.0	0.66		-1.02	-3.84	3.02	0.81
InN[d]	3138	774	1.0	0.68		-6.39	-1.81	8.05	-0.94
AlN[d]	684	298	1.0	1.16		-1.82	-3.43	3.78	0.86
Al$_{0.2}$Ga$_{0.8}$N[d]	306	132	1.0	0.29		-1.33	-1.75	6.02	1.44
In$_{0.2}$Ga$_{0.8}$N[d]	684	386	1.0	1.37		-1.36	-1.95	2.12	-0.99

Holes

Si[a]	460	45	1.23	0.72	−2.18	−0.45		3.2	0.065
Ge[b]	2505	175	1.933	0.9		−0.57	−2.23	2.4	−0.146
GaAs[c]	491	20	1.48	0.38	−2.2			−3.0	
InP[c]	170	10	1.87	0.62	−2.0			−3.0	
GaP[c]	147	10	1.0	0.85	−1.98			−3.0	
AlP[f]	450								
AlAs[c]	200	10	1.84	0.448	−2.24			−3.0	
GaSb[e]	1200	100	1.0	0.5	−2.1				
AlSb[e,f]	460	50	1.0	0.5	−2.2				
InSb[f]	850				−1.95				
InAs[c]	530	20	1.1	0.46	−2.3				
In_{0.53}Ga_{0.47}As[c]	320	10	1.9	0.403	−1.59				
In_{0.49}Ga_{0.51}P[c]	150	15	1.5	0.8	−2.0				
Al_{0.3}Ga_{0.7}As[c]	240	5	1.0	0.324					

Note. Nitride parameters are for the hexagonal structure, all others for cubic crystals.

[a] Reference [60].
[b] Reference [61].
[c] Reference [62].
[d] Reference [63].
[e] Reference [64].
[f] Reference [1].

III–V compounds. If available, mobility measurements should be used to verify these approximations.

Near room temperature ($T = 300$ K) and at low electric fields, the function $\mu(T)$ is often approximated as

$$\mu(T) = \mu_0 \left(\frac{T}{300\,\mathrm{K}} \right)^{\delta_0}. \tag{3.27}$$

Since the mobility is affected by scattering at donors, acceptors, and other defects, it is often desirable to give the mobility as a function of the doping density N_{dop}. The following expression was originally extracted for Si [65] and has later been successfully applied to GaAs [66] and other compounds

$$\mu(N_{\mathrm{dop}}) = \mu_{\mathrm{dop}} + \frac{\mu_0 - \mu_{\mathrm{dop}}}{1 + (N_{\mathrm{dop}}/N_{\mathrm{ref}})^{\alpha}}, \tag{3.28}$$

where μ_0 is the maximum mobility (no doping) and μ_{dop} is the minimum mobility for high doping densities. A temperature dependence of the form (3.27) can be included for any of the parameters μ_0, μ_{dop}, $\mu_1 = \mu_0 - \mu_{\mathrm{dop}}$, N_{ref}, and α [62, 63, 67]. Corresponding exponents δ_0, δ_{dop}, δ_1, δ_{ref}, and δ_{α} are given in Table 3.2, together with the other parameters of Eqs. (3.27) and (3.28). For electrons and holes in GaAs, Eq. (3.28) is plotted in Fig. 3.8 for two different temperatures.

The influence of the electric field F on the mobility can be neglected as long as the carrier drift velocity $v_{\mathrm{dr}} = \mu F$ is much smaller than the Brownian velocity of random motion ($\approx 10^7$ cm/s). In this low-field case, the current density $j_n = qnv_{\mathrm{dr}}$ is proportional to the electric field (Ohm's law). With high fields, the total electron energy of motion becomes large enough to generate longitudinal optical phonons, which eventually leads to a velocity saturation near the Brownian velocity (Fig. 3.9). A commonly used model for the high-field electron mobility $\mu(F)$ in Si has the form

$$\mu(F) = \frac{\mu_{\mathrm{low}}}{\sqrt[\beta]{1 + (\mu_{\mathrm{low}} F/v_s)^{\beta}}}, \tag{3.29}$$

with the low field mobility μ_{low}, the saturation velocity $v_s = 1.1 \times 10^7$ cm/s, and the fit parameter $\beta = 2$ [56]. The same relationship is usually applied to holes in various semiconductors; parameters are listed in Table 3.3. However, the electron mobility in compound semiconductors like GaAs shows a different behavior. Due to electron scattering into side valleys of the conduction band (cf. Fig. 2.6), electron mobility and velocity may exhibit a discrete maximum at larger fields (Fig. 3.9). In this case, the electron mobility is often approximated by the formula

$$\mu_n(F) = \frac{\mu_0 + (v_{sn}/F_0)(F/F_0)^3}{1 + (F/F_0)^4} \tag{3.30}$$

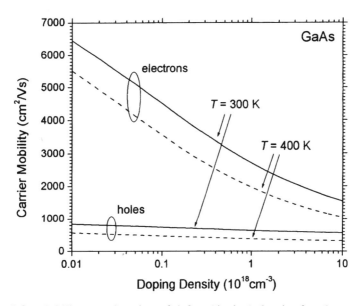

Figure 3.8: Mobility as a function of defect (doping) density for electrons and holes in GaAs at temperatures of 300 (solid) and 400 K (dashed).

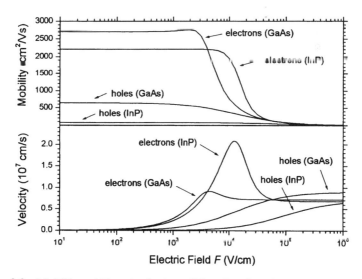

Figure 3.9: Mobility $\mu(F)$ and velocity $\mu(F) \times F$ as functions of the electrostatic field F for electrons (Eq. (3.30)) and holes (Eq. (3.29)) with 10^{18} cm^{-3} doping density ($T = 300$ K).

Table 3.3: Parameters for High-Field Mobility Models (Eqs. (3.29), (3.30), and (3.31)) [4, 56, 68, 69]

Symbol	β	$v_{sp}(300\,\mathrm{K})$	A_s	F_0	$v_{sn}(300\,\mathrm{K})$	A_s
Unit	—	$(10^7\,\mathrm{cm/s})$	—	(kV/cm)	$(10^7\,\mathrm{cm/s})$	—
Si	1.21	0.72	0.37	—	1.02	0.74
Ge	2.0	0.63	0.39	—	0.70	0.45
GaAs	1.0	0.9	0.59	4	0.72	0.44
InP	1.0	0.7		15	0.68	0.31

with the electron saturation velocity v_{sn} and the critical field F_0. The saturation velocity of electrons and holes is temperature dependent and can be written as [68]

$$v_s(T) = \frac{v_s(300\,\mathrm{K})}{(1 - A_s) + A_s T/300\,\mathrm{K}}. \tag{3.31}$$

High-field mobility parameters for some semiconductors are given in Table 3.3; results for other materials are reviewed in [4, 68, 69, 70]. The carrier mobility of nitrides is discussed in Section 9.2.1.

Parallel to surfaces or interfaces, the scattering mechanisms that limit carrier transport are different from bulk mechanisms and special theoretical models have been developed (see, e.g., [53, 56]).

Drift and diffusion of carriers are both limited by the same scatter mechanisms. For parabolic bands and thermal equilibrium, mobility and diffusion coefficient of electrons are connected by [71]

$$D_n = \mu_n \frac{k_B T}{q} \left[\frac{F_{1/2}\left(\frac{E_F - E_c}{k_B T}\right)}{F_{-1/2}\left(\frac{E_F - E_c}{k_B T}\right)} \right] \tag{3.32}$$

$$\approx \mu_n \frac{k_B T}{q} \left[1 + 0.35355 \frac{n}{N_c} - 0.0099 \left(\frac{n}{N_c}\right)^2 + 0.000445 \left(\frac{n}{N_c}\right)^3 \right], \tag{3.33}$$

which simplifies to the well-known Einstein relation $D_n = \mu_n k_B T/q$ for the rather untypical case of nondegenerate semiconductors ($n \ll N_c$). The equivalent relation for holes is obtained by replacing n with p and N_c with N_v.

3.7 Electron–Hole Recombination

The recombination rate R in the continuity equations (3.3) and (3.4) includes various physical mechanisms. In order to drop from the conduction to the valence band, electrons need to transfer their excess energy to other particles (electrons, phonons, photons). The main mechanisms are illustrated in Fig. 3.10. They are radiative (photon emission) or nonradiative (no photon emission).

3.7.1 Radiative Recombination

The generation of photons by radiative recombination can be spontaneous or stimulated. Within bulk layers, the spontaneous emission rate of photons is often characterized by the material coefficient B and written as

$$R_{\text{spon}} = B(np - n_0 p_0). \qquad (3.34)$$

The net recombination rate depends both on the availability of electrons in the conduction band and holes in the valence band. It vanishes in thermal equilibrium with $np = n_0 p_0$. Practically, B is often determined as the second-order coefficient (BN^2) in a polynomial fit to measurements as a function of average carrier concentration N, which results in a wide spreading of reported parameters. A selection

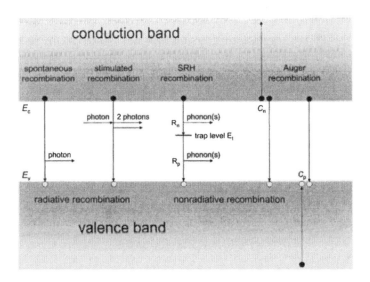

Figure 3.10: Illustration of main electron–hole recombination mechanisms in semiconductors (R_n, R_p, defect recombination rates; C_n, C_p, Auger coefficients).

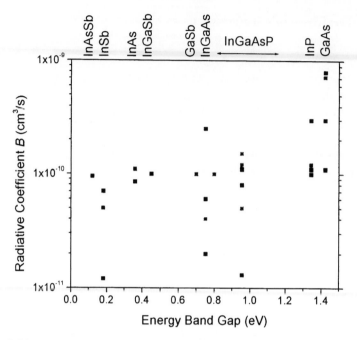

Figure 3.11: Radiative recombination coefficient B for direct band gap compound semiconductors at room temperature [47, 74, 75, 76].

of room-temperature values is given in Fig. 3.11 for various direct-band-gap semi-conductors and their alloys. The parameter B decreases with higher temperature [72] and with tensile strain [73].

Photons can be absorbed in the generation of new electron–hole pairs but they can also trigger recombination and the stimulated emission of another photon. The local stimulated emission rate is proportional to the optical field intensity I_{opt} (W/cm^2)

$$R_{stim} = g \frac{I_{opt}}{\hbar\omega} \tag{3.35}$$

with the optical gain g as a material parameter ($\hbar\omega$, photon energy). Both spontaneous and stimulated emission are of paramount importance in light-emitting optoelectronic devices, and they are further discussed in Chapter 5.

3.7.2 Nonradiative Recombination

The two main nonradiative processes are Shockley–Read–Hall (SRH) recombination and Auger recombination. With Auger recombination, the excess energy is transferred to another electron within the valence or conduction band (Fig. 3.10). Auger recombination may involve different valence bands and the interaction with phonons [77]. The Auger recombination rate is given by

$$R_{\text{Aug}} = (C_n n + C_p p)(np - n_0 p_0),\tag{3.36}$$

with the Auger coefficients C_n and C_p whose temperature dependence is written as

$$C(T) = C_0 \exp\left[-\frac{E_a}{k_B T}\right]\tag{3.37}$$

(E_a, activation energy). Practically, the Auger parameter C is often determined as the third-order coefficient (CN^3) in a polynomial fit to measurements at different carrier concentrations N [72]. Reported room-temperature values are given in Fig. 3.12 as a function of direct band gap energy. Depending on the dominating Auger mechanism, the experimental parameter C represents some combination of the theoretical Auger coefficients C_n and C_p. Auger recombination in long-wavelength InGaAsP quantum wells is further discussed in Chapter 7.

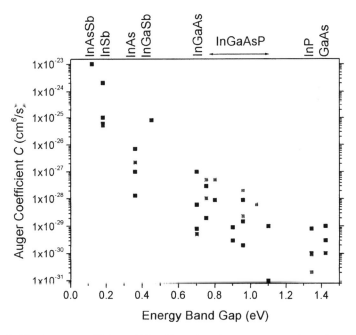

Figure 3.12: Auger coefficient C for direct-band-gap compound semiconductors at room temperature [47, 75, 77].

SRH recombination involves energy levels deep inside the semiconductor band gap that are generated by crystal defects. Such deep-level defects are able to capture electrons from the conduction band as well as holes from the valence band and thereby serve as recombination centers. They are characterized by capture coefficients c_n and c_p, trap density N_t, and trap energy E_t. The capture coefficient is proportional to the capture cross section, which depends on the nature of the defect [2]. Excess electron energy is released by phonons. Based on simple statistics [78, 79], the net electron transition rates from the conduction band R_n and to the valence band R_p is given by

$$R_n = c_n n (1 - f_t) N_t - c_n n_1 f_t N_t \tag{3.38}$$

$$R_p = c_p p f_t N_t - c_p p_1 (1 - f_t) N_t \tag{3.39}$$

with the trap occupancy $0 < f_t < 1$ and

$$n_1 = N_c \exp\left[\frac{E_t - E_c}{k_B T}\right] \tag{3.40}$$

$$p_1 = N_v \exp\left[\frac{E_v - E_t}{k_B T}\right] \tag{3.41}$$

$(n_1 p_1 = n_i^2)$. The dynamic variation of the trap occupancy is described by the difference of the inflow and outflow of electrons

$$N_t \frac{\partial f_t}{\partial t} = R_n - R_p. \tag{3.42}$$

In thermodynamic equilibrium, the net carrier flow through the recombination center vanishes ($R_n = R_p = 0$). Under steady-state conditions, the carrier flow is constant ($R_n = R_p$) and the electron concentration in the trap is given by

$$n_t = f_t N_t = \frac{c_n n + c_p p_1}{c_n (n + n_1) + c_p (p + p_1)} N_t. \tag{3.43}$$

The corresponding net SRH recombination rate is

$$R_{SRH} = N_t \frac{c_n c_p (np - n_0 p_0)}{c_n (n + n_1) + c_p (p + p_1)}. \tag{3.44}$$

The capture coefficients c_n and c_p for electrons and holes relate to SRH lifetimes τ^{SRH} as

$$\frac{1}{\tau_n^{SRH}} = c_n N_t, \tag{3.45}$$

$$\frac{1}{\tau_p^{SRH}} = c_p N_t. \tag{3.46}$$

These SRH lifetimes are typically used as material parameters. For doped semiconductors, Eq. (3.44) can be approximated as

$$R_{\mathrm{SRH}} \approx \frac{p}{\tau_p^{\mathrm{SRH}}} \quad \text{for } p \ll n \tag{3.47}$$

$$R_{\mathrm{SRH}} \approx \frac{n}{\tau_n^{\mathrm{SRH}}} \quad \text{for } n \ll p; \tag{3.48}$$

i.e., SRH recombination is dominated by the more rare recombination partner (minority carrier). Semiconductor materials exhibit different types of defects depending on the fabrication process. Thus, the SRH lifetimes are quite uncertain parameters and should be obtained from fits to measurements. Typical values are in the nanosecond to microsecond range.

Measured carrier lifetimes τ_n and τ_p involve all recombination mechanisms, and they are shorter than SRH lifetimes. For instance, in p-doped material,

$$\frac{1}{\tau_n} \approx \frac{R}{n} = \frac{R_{\mathrm{SRH}} + R_{\mathrm{Aug}} + R_{\mathrm{spon}} + R_{\mathrm{stim}}}{n}$$

$$= \frac{1}{\tau_n^{\mathrm{SRH}}} + \frac{1}{\tau_n^{\mathrm{Aug}}} + \frac{1}{\tau_n^{\mathrm{spon}}} + \frac{1}{\tau_n^{\mathrm{stim}}}. \tag{3.49}$$

3.8 Electron–Hole Generation

Similar to their recombination, the generation of electron–hole pairs requires the interaction with other particles. As illustrated in Fig. 3.13, the transition energy can be provided by phonons (thermal generation), photons (absorption), or other electrons (impact ionization).

The net recombination rates given in the previous section already include thermal generation as they vanish under thermal equilibrium, $np = n_0 p_0$. Thermal ionization of impurity atoms is considered by the neutrality condition of the Poisson equation (3.5). Both thermal generation processes are reflected in the equilibrium Fermi distribution of carriers (cf. Chap. 1).

3.8.1 Photon Absorption

Optical electron–hole pair generation by photon absorption is the key physical mechanism in photodetectors and other electroabsorption devices (see Chapters 11 and 10, respectively). Material aspects of band-to-band absorption are evaluated in Section 4.2.1. It is opposite to stimulated photon emission and included in the net optical gain (see Chap. 5). Negative gain represents net absorption. Due to absorption, the light intensity decreases as the light penetrates deeper into the device.

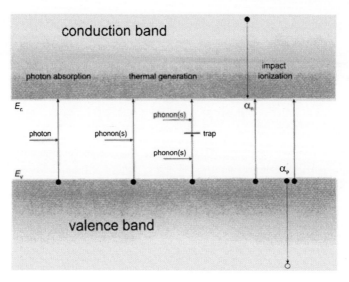

Figure 3.13: Illustration of main electron–hole pair generation mechanisms in semiconductors (α_n, α_p, impact ionization coefficients).

With uniform optical absorption coefficient α_o, the optical generation rate becomes

$$G_o(z) = \alpha_o \frac{I_{\text{opt}}(0)}{\hbar\omega} \exp[-\alpha_o z] \qquad (3.50)$$

with $\hbar\omega$ giving the photon energy and $I_{\text{opt}}(0)$ the optical intensity at the surface (z, penetration distance). The inverse absorption coefficient α_o^{-1} gives the optical penetration depth.

3.8.2 Impact Ionization

Impact ionization is of main importance in devices like avalanche photodetectors, which utilize strong electric fields F and high carrier drift velocities to generate electron–hole pairs. Impact ionization is opposite to Auger recombination as it absorbs the energy of motion of another electron or hole to generate an electron–hole pair. Typically, the impact ionization rate is given as

$$G_{\text{imp}}(F) = \frac{\alpha_n(F)j_n + \alpha_p(F)j_p}{q} \qquad (3.51)$$

with ionization coefficients α_n and α_p for electrons and holes, respectively. The ionization coefficients strongly depend on the electric field (Fig. 3.14), and reported

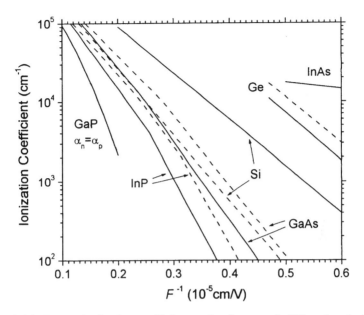

Figure 3.14: Impact ionization coefficient α_n for electrons (solid) and α_p for holes (dashed) vs inverse field strength at room temperature [4, 9, 47].

values vary widely. The field dependence can be approximated by [56]

$$\alpha_{n,p}(F) = \alpha_{n,p}^{\infty} \exp\left[-\left(\frac{F_{n,p}^c}{F} \right)^{\beta_{n,p}^c} \right] \tag{3.52}$$

(n, electrons; p, holes). Typically $\beta^c = 1$ for low fields and $\beta^c = 2$ for high fields. A few of the many reported parameter sets are listed in Table 3.4. In some semiconductors, ionization coefficients are affected by the crystal direction [80].

Various alternative models can be found in the literature, especially for silicon [56]. The following approximation has been adopted for several compound semiconductors (Table 3.5) [81]:

$$\alpha_{n,p}(F) = \alpha_{n,p}^{K} \exp\left[\delta_{n,p} - \sqrt{\delta_{n,p}^2 + (F_{n,p}^K/F)^2} \right]. \tag{3.53}$$

Another, more physics based model for the impact ionization coefficients is described in the following. The mean free flight distance of carriers accelerated

Table 3.4: Impact Ionization Parameters of Eq. (3.52) at Room Temperature

Parameter Unit	Carrier	α^∞ (10^6/cm)	F^c (10^6V/cm)	β^c —	Field range (10^5V/cm)	Ref.
Si	Electrons	0.7	1.23	1	1.75...6	[82]
Si	Holes	1.58	2.04	1	1.75...4	[82]
Si	Holes	0.67	1.7	1	4...6	[82]
Ge ⟨100⟩	Electrons	8.04	1.4	1	1.6...2.1	[83]
Ge ⟨100⟩	Holes	6.39	1.27	1	1.6...2.1	[83]
Ge ⟨111⟩	Electrons	2.72	1.1	1	1.6...2.1	[83]
Ge ⟨111⟩	Holes	1.72	0.937	1	1.6...2.1	[83]
GaAs	Electrons	3.298	1.75	1	3...6	[84]
GaAs	Holes	0.6924	1.384	1	3...6	[84]
InP	Electrons	5.55	3.1	1	3.7...6	[85]
InP	Holes	3.21	2.56	1	3.7...6	[85]

Table 3.5: Impact Ionization Parameters of Eq. (3.53) at Room Temperature [81]

Parameter Unit	Carrier	α^K (10^6/cm)	F^K (10^6V/cm)	δ —	Field range (10^5V/cm)
GaAs	Electrons & holes	0.245	6.65	57.6	3...5
GaP	Electrons & holes	0.39	7.51	19.1	5...13
InP	Electrons	0.55	3.04	2.88	2.2...8
InP	Holes	2.42	3.14	6.8	2.2...8

by the electric field changes with temperature as [57]

$$\lambda_{n,p}(T) = \lambda_{n,p}^0 \tanh\left(\frac{E_{OP}^0}{2k_BT}\right) \tag{3.54}$$

and it is mainly limited by collisions with optical phonons having the average energy

$$E_{OP}(T) = E_{OP}^0 \tanh\left(\frac{E_{OP}^0}{2k_BT}\right). \tag{3.55}$$

Table 3.6: Impact Ionization Parameters for Electrons: High-Field Room-Temperature Mean Free Path λ_n, Low-Temperature Optical Phonon Energy E_{OP}^0, and Ionization Threshold Energy E_n^I [9]

Parameter Unit	λ_n (nm)	E_{OP}^0 (eV)	E_n^I (eV)
Si	6.8	0.063	1.15
Ge	7.8	0.037	0.67
GaAs	3.9	0.035	1.72
InP	3.9	0.043	1.69
GaP	3.2	0.05	2.94
AlP	7.1	0.062	3.14
AlAs	5.4	0.050	2.80
GaSb	4.8	0.024	0.0848
AlSb	4.5	0.042	2.00
InSb	10	0.022	0.187
InAs	6.3	0.030	0.392

The carrier energy gained over the mean free path is $q F \lambda_{n,p}$, and it can be much lower (low-field case) or much higher (high-field case) than the phonon energy. Energy conservation demands that the energy of the ionizing carrier is at least as large as the band gap energy (cf. Fig. 3.13). However, momentum conservation is also required, which leads to a considerably larger ionization threshold energy $E_{n,p}^I$. For isotropic and parabolic bands with equal effective mass, $E_n^I = E_p^I = 3E_g/2$. Reported electron parameters λ_n^0, E_{OP}^0, and E_n^I are listed in Table 3.6. Based on a rigorous numerical model by Baraff [86], analytical fit functions have been derived for the ionization coefficients as functions of field and temperature. The following function is accurate for fields ranging from $E^I/16q\lambda$ to $E^I/5q\lambda$ and $E_{OP}/E^I = 0.01 \ldots 0.06$ [87]

$$\alpha_{n,p}(F, T) = \frac{1}{\lambda_{n,p}} \exp\left[C_0 + C_1 \frac{E_{n,p}^I}{q\lambda_{n,p}F} + C_2 \left(\frac{E_{n,p}^I}{q\lambda_{n,p}F} \right)^2 \right] \tag{3.56}$$

with

$$C_0(T) = -1.92 + 75.5 \frac{E_{OP}}{E_{n,p}^I} - 757 \left(\frac{E_{OP}}{E_{n,p}^I} \right)^2 \tag{3.57}$$

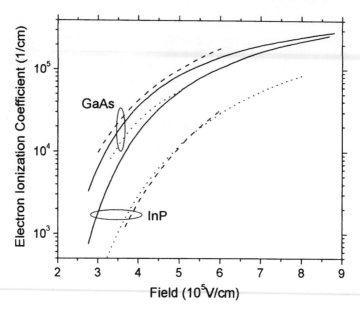

Figure 3.15: Comparison of impact ionization coefficients for electrons in GaAs and InP using the models and parameters given in this section: solid lines, Eq. (3.56) [87]; dashed lines, Eq. (3.56) [56]; dotted lines, Eq. (3.53) [81].

$$C_1(T) = -0.0175 - 11.9 \frac{E_{OP}}{E_{n,p}^I} + 46 \left(\frac{E_{OP}}{E_{n,p}^I} \right)^2 \tag{3.58}$$

$$C_2(T) = 0.00039 - 1.17 \frac{E_{OP}}{E_{n,p}^I} + 11.5 \left(\frac{E_{OP}}{E_{n,p}^I} \right)^2. \tag{3.59}$$

For electrons in GaAs and InP, all three of the above models for the ionization coefficients are compared in Fig. 3.15, illustrating the differences between reported data. These simple impact ionization models assume that the carriers are traveling in a constant electric field. This is often not the case in real devices, and the ionization rates are typically overestimated by above formulas. For a more detailed discussion of impact ionization, see [9, 53, 56, 57, 88, 89].

3.8.3 Band-to-Band Tunneling

Finally, carriers can be generated without additional energy, by band-to-band tunneling with strong electric fields $F > 10^6$ V/cm (Fig. 3.16). The corresponding

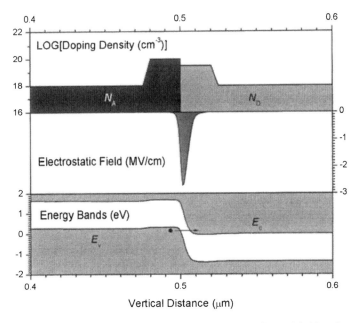

Figure 3.16: Illustration of band-to-band tunneling in a highly doped InP *pn*-junction at 0.5 V reverse bias.

generation rate is often calculated using [90, 91]

$$G_{bbt}(F) = A_{bbt} F^{\gamma_{bbt}} \exp\left[-\frac{B_{bbt}}{F}\right] \tag{3.60}$$

with $\gamma_{bbt} = 2.5$, $A_{bbt} = 4 \times 10^{14}\,\text{cm}^{-0.5}\,\text{V}^{-2.5}\,\text{s}^{-1}$, and $B_{bbt} = 1.9 \times 10^7\,\text{V/cm}$ for silicon (phonon-assisted indirect transition). For tunneling in direct-band-gap semiconductors, $\gamma_{bbt} = 2$ and

$$A_{bbt} = \frac{\sqrt{2m_{bbt}}q^2}{\pi h^2 E_g^{1/2}} \tag{3.61}$$

$$B_{bbt} = \frac{\pi^2 \sqrt{m_{bbt}} E_g^{3/2}}{\sqrt{2}qh}, \tag{3.62}$$

where the effective tunneling mass is an average of electron and hole mass

$$m_{bbt} = \frac{2m_c m_v}{m_c + m_v}. \tag{3.63}$$

This simple band-to-band tunneling model assumes a constant electrical field accross the tunnel junction, which may be not the case (cf. Fig. 3.16). Accurate treatment of tunneling in real devices requires more sophisticated numerical methods (cf. Section 3.4).

3.9 Advanced Transport Models

Thus far, this chapter has focused on the widely used drift–diffusion model for the carrier transport in semiconductor devices. However, more advanced transport models have been developed, some of which are briefly outlined in the following. For simplicity, the crystal lattice temperature T_L is assumed uniform and constant in this section. Lattice heating effects are discussed in Chap. 6.

3.9.1 Energy Balance Model

In thermal equilibrium, the average energy of motion per carrier equals $3k_B T_L/2$. In order to include nonequilibrium carrier distributions with higher average energy, the carrier temperatures T_n and T_p are introduced as new independent parameters for electrons and holes, respectively. In thermal equilibrium, T_n and T_p are equal to the lattice temperature T_L. With the carrier energy as new unknown parameter, energy balance equations have been developed in addition to the carrier balance equations in the drift–diffusion model [92, 93, 94].

Several new quantities are used in the energy balance model. Multiplying the average carrier energy by the carrier concentration gives the carrier energy densities

$$w_n = \frac{3}{2}k_B T_n n \tag{3.64}$$

$$w_p = \frac{3}{2}k_B T_p p, \tag{3.65}$$

which are functions of local position $\vec{r}$ and time t. The flow of carrier energy can be independent from the flow of carriers. The carrier energy flux densities are given by

$$\vec{S}_n = -\kappa_n \nabla T_n + \pi_n \vec{j}_n \tag{3.66}$$

$$\vec{S}_p = -\kappa_p \nabla T_p + \pi_p \vec{j}_p, \tag{3.67}$$

representing conductive (first term) and convective (second term) heat transport by carriers. The latter is also known as the Peltier effect with the Peltier coefficients π_n and π_p for electrons and holes, respectively. The carrier thermal conductivities

κ_n and κ_p are related to the mobilities by the Wiedemann–Franz law

$$\kappa_n = \frac{3k_{\mathrm{B}}^2}{2q}\mu_n n T_n \tag{3.68}$$

$$\kappa_p = \frac{3k_{\mathrm{B}}^2}{2q}\mu_p p T_p. \tag{3.69}$$

Carriers always travel in a direction that allows for lower carrier energy. Besides drift and diffusion, the energy balance model includes additional driving forces for the flow of carriers, which lead to the current densities

$$\vec{j}_n = \mu_n \left[n\nabla E_{\mathrm{c}} + k_{\mathrm{B}} T_n \nabla n - q P_n n \nabla T_n - w_n \nabla \ln(m_{\mathrm{c}}) \right] \tag{3.70}$$

$$\vec{j}_p = \mu_p \left[p\nabla E_{\mathrm{v}} - k_{\mathrm{B}} T_p \nabla p - q P_p p \nabla T_p - w_p \nabla \ln(m_{\mathrm{v}}) \right]. \tag{3.71}$$

The first term describes variations of the band edges as the driving force, including the electrostatic potential (drift) as well as variations of material composition and band gap (Fig. 3.17a). The common diffusion term is second, employing the Einstein relation for the diffusion coefficients. Carrier temperature variations

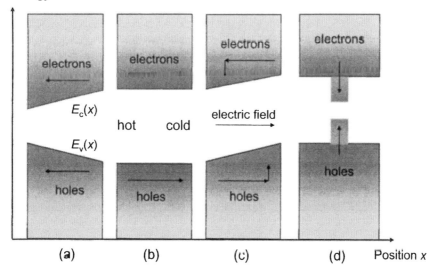

Figure 3.17: Illustration of transport effects described by the energy balance model: (a) current driven by band gap variation, (b) current driven by temperature gradient, (c) Joule heat generation, and (d) hot carrier capture by quantum wells.

are the driving force in the third term. The generation of current by temperature gradients ∇T is called the Seebeck effect with the thermoelectric powers P_n and P_p (V/K), respectively, as material parameter. The thermoelectric power accounts for the extra energy of carriers above the Fermi level. This energy increases with higher temperature due to the wider spreading of the Fermi function. When a temperature gradient occurs, carriers move from hot regions to cold regions in order to reduce that extra energy (Fig. 3.17b). For nondegenerate semiconductors, the approximation

$$P_n = -\frac{k_B}{q}\left[r + \ln\left(\frac{N_c}{n}\right)\right] \tag{3.72}$$

$$P_p = \frac{k_B}{q}\left[r + \ln\left(\frac{N_v}{p}\right)\right] \tag{3.73}$$

is often used, where r depends on the dominant carrier scatter mechanism [2]:

$r = 1$ for amorphous semiconductors
$r = 2$ for acoustic phonon scattering
$r = 3$ for optical phonon scattering
$r = 4$ for ionized impurity scattering
$r = 2.5$ for neutral impurity scattering.

The thermoelectric power is related to the Peltier coefficient by the Kelvin relation

$$\pi_n = P_n T_n \tag{3.74}$$

$$\pi_p = P_p T_p. \tag{3.75}$$

The last term in Eqs. (3.70) and (3.71) accounts for spatial variations of the effective mass as the driving force for the carrier transport.

In the drift–diffusion model, Eqs. (3.3) and (3.4) represent a conservation law for carriers. The energy conservation law is represented by the energy balance equations

$$\frac{\partial w_n}{\partial t} + \nabla \cdot \vec{S}_n = -\frac{3}{2}k_B T_n (R - G) + \vec{j}_n \cdot \nabla E_c - \frac{3}{2}k_B n \frac{T_n - T_L}{\tau_{E,n}} \tag{3.76}$$

$$\frac{\partial w_p}{\partial t} + \nabla \cdot \vec{S}_p = -\frac{3}{2}k_B T_p (R - G) - \vec{j}_p \cdot \nabla E_v - \frac{3}{2}k_B p \frac{T_p - T_L}{\tau_{E,p}}. \tag{3.77}$$

With constant total energy of electrons, the local change of their energy density $\partial w_n / \partial t$ must be compensated for by the net flow of electron energy $\nabla \cdot \vec{S}_n$ at that location (same for holes). Thus, the right-hand side (RHS) of these equations is zero as long as no carrier energy is gained or lost. Energy loss or gain can be related to the recombination or generation of electrons and holes (first term on RHS). Energy gain can also be caused by Joule heating, as given in the second RHS term. Without scattering, the current flow toward lower band energy separates the carriers from the band edge as they maintain their energy (Fig. 3.17c). Carrier scattering eventually transforms this energy gain into lattice heating. Such relaxation of electrons from higher to lower band energies is described by the third RHS term with the energy relaxation time τ_E. The relaxation times are the key material parameters of the energy balance model, as they determine how quickly carriers reach an equilibrium Fermi distribution with $T_n = T_p = T_L$. This is of special importance for quantum wells, for instance, where electrons entering the well suddenly have too much energy (hot electrons), which needs to be dissipated in order to populate quantum levels (Fig. 3.17d). The parameter τ_E is hardly measurable, and it is often calculated by Monte Carlo analysis [95]. The relaxation time τ_E, the mobility μ_n and other parameters depend on the energy or temperature of carriers. In temperature-dependent formulas for the effective density of states and other quantities, the lattice temperature is replaced by the corresponding carrier temperature.

3.9.2   Boltzmann Transport Equation

Drift–diffusion and energy balance model are both simplified versions of a more general treatment of carrier transport. It employs the general electron distribution function $f(\vec{r}, \vec{k}, t)$, which can be understood as a combination of local distribution $n(\vec{r}, t)$ and energy distribution $f(E, t)$. The Boltzmann transport equation gives the time derivative

$$\frac{df}{dt} = \frac{\partial f}{\partial t} + \frac{\partial \vec{k}}{\partial t}\nabla_k f + \frac{\partial \vec{r}}{\partial t}\nabla f = -\left(\frac{\partial f}{\partial t}\right)_{coll}. \tag{3.78}$$

The first term of the middle part represents the direct time dependence of f; it is zero under steady-state conditions. The second term includes carrier momentum changes due to the electric field ($\partial \vec{k}/\partial t = -q\vec{F}/\hbar$, ∇_k describes the derivative in $\vec{k}$ space). The third term includes the carrier velocity $\partial \vec{r}/\partial t = \vec{v}$; it vanishes for homogeneous semiconductors ($\nabla f = 0$). The by-far most complicated part of the Boltzmann equation is the right-hand side, which integrates all possible collision mechanisms. It includes any change of momentum $\hbar \vec{k}$ or energy E due to scatter events, both elastic (E constant) and inelastic (E changed). The RHS of the Boltzmann equation vanishes for carrier equilibrium when f is identical to the Fermi distribution function.

Finding solutions to this complicated integro-differential equation has been a challenge for countless researchers over decades. Among the direct methods of solution, the most important ones are the Monte Carlo method, which numerically follows the path of single carriers [95], and the relaxation time approximation, which replaces the collision term by the simple approximation

$$\left(\frac{\partial f}{\partial t}\right)_{coll} = \frac{f - f_0}{\tau_m} \tag{3.79}$$

employing a momentum relaxation time τ_m to describe how quickly f assumes an equilibrium distribution f_0 after all external forces are turned off. In this approach, τ_m is the only material parameter, leading to a more general definition of the carrier mobility and other material parameters [89].

Drift–diffusion and energy balance model can be derived from the Boltzmann transport equation by the method of moments. This method multiplies the entire equation by a specific quantity and then averages over the entire k space [53]. It leads to conservation laws for the carrier concentration (zeroth moment, Eq. (3.3)), for the current density (first moment, Eqs. (3.70) or (3.1)), and for the carrier energy density (second moment, Eq. (3.77)). Besides the energy balance model (Section 3.9.1), a variety of more sophisticated carrier transport models has been established this way, including the hydrodynamic model [96] and the energy transport model [97].

Further Reading

- S. Selberherr, *Simulation and Analysis of Semiconductor Devices*, Springer-Verlag, New York, 1984.

- K. Hess, *Advanced Theory of Semiconductor Devices*, IEEE Press, Piscataway, NJ, 2000.

- N. G. Einspruch and W. R. Frensley, *Heterostructures and Quantum Devices*, Academic Press, San Diego, 1994.

Chapter 4

Optical Waves

Optical waves are combinations of high-frequency electrical fields $\vec{E}$ and magnetic fields $\vec{H}$ in the wavelength range from the infrared through the visible to the ultraviolet light (Fig. 4.1). The mathematical treatment of optical waves is embedded in the classical theory of electromagnetic fields based on Maxwell's equations. These general equations lead to the Helmholtz wave equations, which govern wave optics. As the intensity distribution of the optical wave is of main interest, scalar approximations of those equations are often employed. Ray optics represent a further simplification for optical structures much larger than the wavelength. Entire books are devoted to optical wave theory, and a great variety of numerical methods for solving the Helmholtz equations within wave guiding structures has been developed. This chapter gives a basic introduction to optical wave theory, focusing on semiconductor optoelectronic device issues.

4.1 Maxwell's Equations

Electromagnetic fields in general are governed by Maxwell's equations

$$\nabla \times \vec{E} = -\frac{\partial}{\partial t} \mu \mu_0 \vec{H} \tag{4.1}$$

$$\nabla \times \vec{H} = \vec{j} + \frac{\partial}{\partial t} \varepsilon \varepsilon_0 \vec{E} \tag{4.2}$$

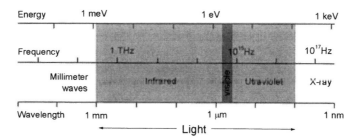

Figure 4.1: Electromagnetic spectrum indicating the optical wavelength window.

$$\nabla \cdot \varepsilon\varepsilon_0 \vec{E} = \varrho \tag{4.3}$$

$$\nabla \cdot \mu\mu_0 \vec{H} = 0 \tag{4.4}$$

with the electrical permittivity $\varepsilon\varepsilon_0$ and the magnetic permeability $\mu\mu_0$ describing the interaction of the material with electrical and magnetic fields, respectively, in linear approximation (ε_0, μ_0, free-space values; $\vec{j}$, current density; ϱ, charge density). Maxwell's equations are the fundamental basis for the classical treatment of electrical and magnetic fields. For high-frequency optical waves and semiconductor waveguides, several simplifications are possible as discussed in the following.

The charge density ϱ acts as source of the electrostatic field $\vec{F}$, and it can be neglected in the case of high-frequency optical fields $\vec{E}$ (cf. Poisson equation (3.5)). Thus, the source-free divergence equations (4.3) and (4.4) indicate that both the electrical and the magnetic parts of the optical field are closed curls and do not originate from single sources. Magnetic field variations in time $\partial \vec{H}/\partial t$ generate curls $\nabla \times \vec{E}$ of the electrical field (Eq. (4.1)). Vice versa, electrical field variations in time $\partial \vec{E}/\partial t$ as well as the current $\vec{j}$ generate curls $\nabla \times \vec{H}$ of the magnetic field (Eq. (4.2)). The steady-state or low-frequency current density $\vec{j} = \sigma \vec{F}$ can be excluded here (σ, conductivity at low frequencies). The remaining "optical" current density $\vec{j} = \sigma_{opt}\vec{E}$ depends on the semiconductor conductivity σ_{opt} at high frequencies. Even without any carriers present, the wavelike change of $\vec{E}(\vec{r}, t)$ and $\vec{H}(\vec{r}, t)$ constantly generates the other field, thereby keeping the electromagnetic wave alive (free space: $\varepsilon = 1$, $\mu = 1$, $\vec{j} = 0$, $\varrho = 0$).

The interaction of the magnetic field vector $\vec{H}$ with semiconductors is typically weak, and it will be neglected in the following ($\mu = 1$). The electrical field $\vec{E}$ interacts with free and bound charges in semiconductors. Under the influence of the electrical field, bound charges (valence electrons and ionic atom cores) are slightly separated from their average position, thereby creating an internal electrical polarization

$$\vec{P} = \varepsilon_0 \chi \vec{E} \tag{4.5}$$

with the electrical susceptibility χ as material parameter, which is related to the dielectric constant by $\varepsilon = \chi + 1$. In anisotropic materials, the susceptibility χ is a matrix and the polarization field $\vec{P}$ points in a different direction than $\vec{E}$. However, most semiconductors of interest here belong to the cubic crystal system and can be considered isotropic.[1] Furthermore, the polarization depends on the frequency of the electric field. At low frequencies and long wavelengths, all charges are able to respond to the field variation. At higher frequencies and shorter wavelengths (below $\lambda \approx 10\,\mu m$), the more heavy ionic atom cores

---

[1] For a treatment of anisotropic properties, see, e.g., [98].

cannot follow the field variations any more. Thus, the semiconductor's response to optical fields is mainly governed by electrons in the valence and conduction bands. The values of material parameters at optical frequencies are different from static parameters.[2] Whereas the static dielectric constant ε_{st} is used for static and low-frequency fields $\vec{F}$, the high-frequency value ε_{opt} is employed for optical fields $\vec{E}$.

Assuming monochromatic radiation and time-harmonic field variations $\propto \exp(i\omega t)$ with the angular frequency $\omega = 2\pi\nu$, we obtain the semiconductor Maxwell equations in the frequency domain

$$\nabla \times \vec{E} = -i\omega\mu_0\vec{H} \tag{4.6}$$

$$\nabla \times \vec{H} = (\sigma_{opt} + i\omega\varepsilon_{opt}\varepsilon_0)\vec{E} \tag{4.7}$$

$$\nabla \cdot \varepsilon_{opt}\varepsilon_0\vec{E} = 0 \tag{4.8}$$

$$\nabla \cdot \mu_0\vec{H} = 0, \tag{4.9}$$

where $\vec{E}$ and $\vec{H}$ were changed to the time-independent fields $\vec{E}$ and $\vec{H}$, respectively. The term $(\sigma_{opt} + i\omega\varepsilon_{opt}\varepsilon_0)$ suggests a complex number generalization of the dielectric constant.

4.2 Dielectric Function

The dielectric function $\varepsilon(\omega)$ describes the response of a material to an electrical field $\vec{E}$. This response depends on the angular frequency ω of the field so that its value for high-frequency optical fields (ε_{opt}) is different from the value at low-frequency or static fields (ε_{st}). In order to include absorption, the dielectric function is written in complex notation

$$\tilde{\varepsilon}(\omega) = \varepsilon'(\omega) + i\varepsilon''(\omega) \tag{4.10}$$

with the real part ε' describing polarization and the imaginary part ε'' describing gain ($\varepsilon'' > 0$) or absorption ($\varepsilon'' < 0$). Note that complex numbers are an artificial construction for mathematical convenience. Any real physical quantity is represented by a real number, which can be the real part ($\varepsilon' = \text{Re}\{\tilde{\varepsilon}\}$) or the imaginary part ($\varepsilon'' = \text{Im}\{\tilde{\varepsilon}\}$) of the complex number.

Instead of $\tilde{\varepsilon}$, the complex refractive index

$$\tilde{n}(\omega) = n_r(\omega) + ik_a(\omega) \tag{4.11}$$

---

[2]Exceptions are nonpolar semiconductors like Si and Ge.

is often used with $\tilde{n}^2 = \tilde{\varepsilon}$. The (real) refractive index n_r reduces the free-space light velocity c_0 to

$$c = \frac{c_0}{n_r} \qquad (4.12)$$

inside the material. The extinction ratio

$$k_a = \frac{\lambda \alpha_0}{4\pi} \qquad (4.13)$$

is related to the optical power absorption coefficient α_0. The influence of both parts on an electromagnetic wave is illustrated in Fig. 4.2 for the electric field

$$\vec{E}(z) \propto \exp\left[i\frac{\omega}{c}z\right] = \exp\left[i\frac{2\pi\nu}{c_0}\tilde{n}z\right] = \exp\left[-\frac{\alpha_0}{2}z\right] \times \exp\left[i\frac{2\pi}{\lambda_0}n_r z\right]. \quad (4.14)$$

The real part n_r reduces the free-space wavelength to

$$\lambda = \frac{\lambda_0}{n_r} \qquad (4.15)$$

whereas the absorption coefficient α_0 reduces the wave's amplitude.

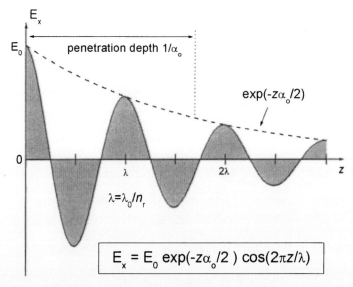

Figure 4.2: Illustration of the electrical field of an optical wave within lossy material (Eq. (4.14)).

A set of only two material parameters fully describes the interaction between optical radiation and semiconductor materials: n_r and k_a (α_o), or ε' and ε''. The relation of the two sets is given by

$$\varepsilon' = n_r^2 - k_a^2 \tag{4.16}$$

$$\varepsilon'' = 2n_r k_a \tag{4.17}$$

$$n_r^2 = \frac{1}{2}\left[\varepsilon' + \sqrt{\varepsilon'^2 + \varepsilon''^2}\right] \tag{4.18}$$

$$k_a^2 = \frac{1}{2}\left[-\varepsilon' + \sqrt{\varepsilon'^2 + \varepsilon''^2}\right]. \tag{4.19}$$

An alternative third set of optical parameters is the complex susceptibility

$$\tilde{\chi} = \tilde{\varepsilon} - 1. \tag{4.20}$$

In all three cases, both parameters are related by the Kramers–Krönig equations, for instance,

$$\varepsilon'(\omega_0) - 1 = \frac{2}{\pi}\mathbf{P}\int_0^\infty d\omega\, \frac{\omega\varepsilon''(\omega)}{\omega^2 - \omega_0^2} \tag{4.21}$$

$$\varepsilon''(\omega_0) = \frac{2\omega_0}{\pi}\mathbf{P}\int_0^\infty d\omega\, \frac{\varepsilon'(\omega) - 1}{\omega_0^2 - \omega^2} \tag{4.22}$$

(**P** indicates a Cauchy principal integral that excludes the singular point $\omega = \omega_0$) [20]. The Kramers–Krönig relations permit us to determine either the real or the imaginary component if the other is known for all frequencies. For example, knowledge of the full absorption spectrum $\alpha_o(\omega)$ allows for the calculation of the refractive index n_r at any frequency ω_0. Without any absorption, $n_r \equiv 1$ at all frequencies. An absorption peak at a given wavelength most strongly affects the refractive index close to that wavelength. For GaAs, the spectra $\alpha_o(\lambda_0)$ and $n_r(\lambda_0)$ are plotted in Fig. 4.3. The details are discussed in the following. Many theoretical material models analyze the relevant absorption mechanisms first and then derive the refractive index using the Kramers–Krönig relation.

4.2.1 Absorption Coefficient

Band-to-band electron transitions from the valence to the conduction band constitute the main absorption process in optoelectronic devices such as photodetectors and electroabsorption modulators. In direct semiconductors like GaAs and InP, the absorption for photon energies $\hbar\omega$ above the band gap E_g can be approximated as

$$\alpha_o \propto \sqrt{\hbar\omega - E_g}. \tag{4.23}$$

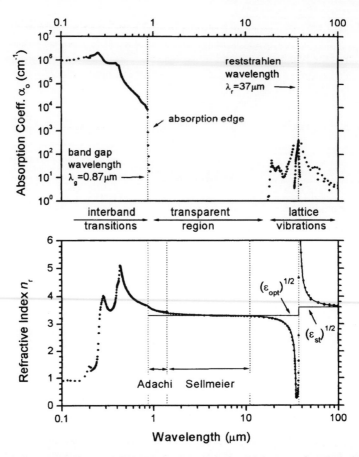

Figure 4.3: Absorption coefficient α_0 and refractive index n_r of undoped GaAs at room temperature as a function of wavelength. The validity range of Eq. (4.31) (simplified Adachi model) and of Eq. (4.28) (Sellmeier model), as well as for the static dielectric constant ε_{st} and for the optical dielectric constant ε_{opt} is indicated (measured data points are from [99]).

For indirect semiconductors like Si and Ge, phonons need to be involved to provide momentum, and the band-to-band absorption coefficient can be described as

$$\alpha_0 \propto (\hbar\omega \pm E_P - E_g)^2 \tag{4.24}$$

with the phonon energy E_P (plus/minus stands for phonon absorption/generation). Additional bands with slightly higher gap energy may also be important (like the

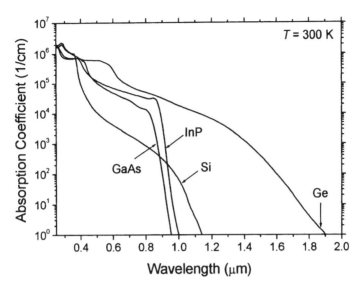

Figure 4.4: Measured band-to-band absorption coefficient versus wavelength for different undoped semiconductors [99]. The arrows mark the band gap wavelength.

spin–orbit valence band in III–V compounds). Interband absorption spectra for several semiconductors are given in Fig. 4.4. As the bands fill up with carriers, the optical absorption edge rises above the band gap E_g (Moss–Burstein shift), which is partially compensated for by band gap reduction with higher carrier concentration N (see Section 2.1). For the optical band gap, a Moss–Burstein correction of $\Delta E_g = 1.6 \times 10^{-8} N^{1/3}$ was given for InP-based compounds [100]. The band gap E_g is also affected by temperature and strain (see Chap. 2). Band-to-band transitions are further discussed in Chap. 5.

For photon energies slightly below the optical band gap, the absorption coefficient was found to change exponentially as

$$\alpha_0 = A_{\text{tail}} \exp\left[\frac{\hbar\omega - E_g}{E_{\text{tail}}}\right] \tag{4.25}$$

with material parameters $A_{\text{tail}} = 3000/\text{cm}$ and $E_{\text{tail}} = 10\,\text{meV}$ for GaAs and InP [100]. E_{tail} is also referred to as the Urbach parameter, and it accounts for band tails in doped or disordered semiconductors [69].

As the band-to-band absorption decays with lower photon energy, transitions within conduction and valence band(s), respectively, emerge as the dominant absorption mechanism. These are often called free-carrier absorption; however,

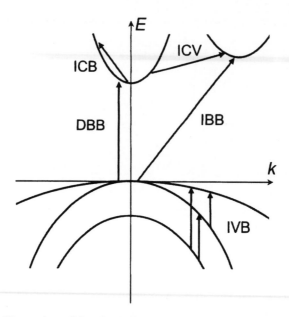

Figure 4.5: Illustration of the absorption processes within a typical energy band structure $E(k)$: DBB, direct band-to-band absorption; IBB, indirect band-to-band absorption; IVB, intervalence band absorption; ICB, intraconduction band absorption; ICV, interconduction valley absorption.

different types of transitions can be involved (Fig. 4.5). In general, one distinguishes *intraband* transitions within the same band, involving phonons, and *interband* transitions between different bands. The free-carrier absorption coefficient can be approximated as a linear function of the carrier concentration

$$\alpha_0 = k_n n + k_p p. \tag{4.26}$$

In InP and related compounds, intervalence band absorption (IVBA) is of special importance within the fiber optic wavelength window (1.3–1.6 μm) as the energy separation between valence bands matches the photon energy. The IVBA coefficient was found to be very similar in p-doped InP, GaAs, and $In_{0.53}Ga_{0.47}As$ with $k_p = 13 \times 10^{-18}$ cm^2 at 1.3 μm and 25×10^{-18} cm^2 at 1.6 μm [101]. Transitions within the conduction band cause much weaker absorption in n-doped material; $k_n = 2 \times 10^{-18}$ cm^2 was extracted at 1.3 μm for $In_{1-x}Ga_xAs_yP_{1-y}$ lattice-matched to InP ($y = 0 \dots 0.4$) [100]. Calculations of free-carrier absorption need to consider the nonparabolic band structure [102]. The free-carrier intraband absorption coefficient can be related to the high-frequency electrical conductivity σ_{opt} by

$$\alpha_0 = \frac{\sigma_{opt}}{\varepsilon_0 c_0 n_r}, \tag{4.27}$$

and it is proportional to λ^p with $p = 2 \ldots 3$ [103].

At even lower photon energies, absorption is dominated by interactions with the crystal lattice (*reststrahlen* region, see Fig. 4.3). However, those wavelengths are usually not of interest in optoelectronics.

4.2.2 Index of Refraction

For photon energies below the band gap, empirical oscillator models are often used to describe the wavelength (energy) dependence of the refractive index. Those models reflect the fact that optical waves interact with charged particles that behave like mechanical oscillators. However, the oscillator frequencies used are often pure fit parameters. One of the most popular models for dielectric materials is the second-order Sellmeier formula

$$n_r^2(\lambda) = s_0 + \frac{s_1 \lambda^2}{\lambda^2 - \lambda_1^2} + \frac{s_2 \lambda^2}{\lambda^2 - \lambda_2^2}, \tag{4.28}$$

which uses fit parameters s_i and λ_i as listed in Table 4.1 [104]. The parameters λ_1 and λ_2 are the oscillator's resonance wavelengths, which are placed outside

Table 4.1: Parameters s_i and λ_i of the Sellmeier Refractive Index Model for Undoped Semiconductors at Room Temperature (Eq. (4.28)) [104]

Parameter Unit	s_0	s_1	λ_1 (μm)	s_2	λ_2 (μm)	Range (μm)
Si	3.129	8.54279	0.33671	0.00528	38.72983	1.36–11
Ge	9.282	6.7288	0.66412	0.21307	62.20932	2–12
GaAs	3.5	7.4969	0.4082	1.9347	37.17526	1.4–11
InP	7.255	2.316	0.6263	2.765	32.93934	0.95–10
AlAs	2.616	5.56711	0.2907	0.49252	10	0.56–2.2
GaSb	13.1	0.75464	1.2677	0.68245	10	1.8–2.5
InAs	11.1	0.71	2.551	2.75	45.6618	3.7–31.3
GaP	3.096	5.99865	0.30725	0.83878	17.32051	0.54–4.0
InSb	15.4	0.10425	7.15437	3.47475	44.72136	7.8–22
GaN	3.6	1.75	0.256	4.1	17.86057	<10
AlN	3.14	1.3786	0.1715	3.861	15.0333	0.22–5.0

the range of validity. The result for GaAs is shown in Fig. 4.3. Sellmeier coefficients can be interpolated to find formulas for multinary compounds (InGaAsP [100], AlGaInAs [105], AlGaInP [106]). A single-effective-oscillator model was proposed by Wemple and DiDomenico [107], giving

$$n_r^2(\omega) = 1 + \frac{E_o E_d}{E_o^2 - \hbar\omega} \tag{4.29}$$

with the oscillator energy E_o and the dispersion energy E_d as adjustable parameters. This model has also been used for some multinary III–V compounds [108, 109]. However, both of these phenomenological models tend to underestimate the refractive index at photon energies close to the semiconductor band gap [100]. There, the refractive index is mainly influenced by band-to-band absorption processes. Accordingly, Afromowitz has modified the single-oscillator model to include absorption at the band gap energy E_g,

$$n_r^2(\omega) = 1 + \frac{E_d}{E_o} + (\hbar\omega)^2 \frac{E_d}{E_o^3}$$

$$+ (\hbar\omega)^4 \frac{E_d}{2E_o^3(E_o^2 - E_g^2)} \times \ln\left[\frac{2E_o^2 - E_g^2 - (\hbar\omega)^2}{E_g^2 - (\hbar\omega)^2}\right], \tag{4.30}$$

which gives excellent agreement with measurements on AlGaAs and GaAsP [110]. This model has also been used for GaInP [110] and GaInAsP [111].

Adachi developed a more physics-based model of the dielectric function for photon energies close to and above the band gap E_g [69]. The model can be simplified for the transparency region ($\hbar\omega < E_g$) by considering only the first two band gaps (E_g, $E_g + \Delta_0$)

$$n_r^2(\omega) = A\left[f(x_1) + 0.5\left(\frac{E_g}{E_g + \Delta_0}\right)^{1.5} f(x_2)\right] + B \tag{4.31}$$

with

$$f(x_1) = \frac{1}{x_1^2}\left(2 - \sqrt{1 + x_1} - \sqrt{1 - x_1}\right), \quad x_1 = \frac{\hbar\omega}{E_g} \tag{4.32}$$

$$f(x_2) = \frac{1}{x_2^2}\left(2 - \sqrt{1 + x_2} - \sqrt{1 - x_2}\right), \quad x_2 = \frac{\hbar\omega}{E_g + \Delta_0} \tag{4.33}$$

and is shown to give good agreement with measurements on InGaAsP by linear interpolation of the binary material parameters A and B [69]. Both parameters are listed in Table 4.2 for various binary compounds [43]. In nitride III–V compounds,

Table 4.2: Parameters for the Simplified Adachi Model for the Refractive Index below the Band Gap ($\hbar\omega < E_g$) as Given in Eqs. (4.31) [43] and (4.34) [112, 120]

Parameter	A	B	E_g	Δ_0
GaAs	6.30	9.40	1.42	0.34
InP	8.40	6.60	1.35	0.11
AlAs	25.30	−0.80	2.95	0.28
GaSb	4.05	12.66	0.72	0.82
AlSb	59.68	−9.53	2.22	0.65
InAs	5.14	10.15	0.36	0.38
GaP	22.25	0.90	2.74	0.08
AlP	24.10	−2.00	3.58	0.07
InSb	7.91	13.07	0.17	0.81
GaN	9.84	2.74	3.42	—
AlN	13.55	2.05	6.28	—
InN	53.57	−9.19	1.89	—

the valence band splitting is very small and Adachi's model for the transparency region can be reduced to only one interband transition at E_g giving

$$n_r^2(\omega) = A\left(\frac{\hbar\omega}{E_g}\right)^{-2}\left\{2 - \sqrt{1 + \left(\frac{\hbar\omega}{E_g}\right)} - \sqrt{1 - \left(\frac{\hbar\omega}{E_g}\right)}\right\} + B \qquad (4.34)$$

and showing good agreement with measurements on GaN, AlN, and InN [112] (see Section 9.2.4 for nitride alloys). Adachi's refractive index model has been applied to Si and Ge [113] as well as to many III–V compounds used in optoelectronics (AlGaAs [114]; InGaAsP [115]; AlGaAsSb, GaInAsSb, and InPAsSb [116]; AlGaInP [117]; ZnCdTe [118]; AlGaInN [119]). However, for specific materials and selected wavelength regions, tailored models may show better agreement with measurements (see, e.g., [100] for InGaAsP with $\hbar\omega \approx E_g$).

For lower energies near the transversal optical phonon energy, the refractive index is affected by photon interaction with the crystal lattice [2]. The phonon energy translates into the *reststrahlen* wavelength, which separates the optical wavelength region ($n_r \approx \sqrt{\varepsilon_{opt}}$) from the low-frequency region ($n_r \approx \sqrt{\varepsilon_{st}}$) as illustrated in Fig. 4.3. All three parameters are listed in Table 4.3.

The effect of the free carrier concentration on the refractive index is related to band gap renormalization, band filling, and free-carrier absorption.

Table 4.3: Static (ε_{st}) and Optical (ε_{opt}) Dielectric Constants, *Reststrahlen* Wavelength λ_r [99], Band Gap Wavelength λ_g, Refractive Index n_r at Band Gap Wavelength, and Refractive Index Change with Temperature

| Parameter | ε_{st} | ε_{opt} | λ_r | λ_g | $n_r(\lambda_g)$ | $(dn_r/dT)/n_r$ |
Unit	—	—	(μm)	(μm)	—	(10^{-5} K^{-1})
GaAs	12.91	10.9	37	0.87	3.65	4.5
InP	12.61	9.61	32	0.92	3.41	2.7
AlAs	10.06	8.16		0.42	3.18	4.6
GaSb	15.69	14.44		1.72	3.82	8.2
AlSb	12.04	10.24		0.56	3.40	3.5
InAs	15.15	12.25	45	3.45	3.52	6.5
GaP	11.10	9.08		0.45	3.45	2.5
AlP	9.80	7.54		0.35	3.03	3.5
InSb	17.70	15.68	56	7.3	4.00	12.0
GaN	8.9	5.35		0.36	2.67	
AlN	8.5	4.6		0.20	2.5	
InN	15.3	8.4		0.65	3.15	

Note. Data are from [5] except for the wurtzite nitrides [38]. Note that $n_r^2(\lambda_g)$ is larger than ε_{opt}.

It has been investigated in [121] for InP, GaAs, and InGaAsP. Refractive index changes with temperature are mainly based on the dependence $E_g(T)$ and are often considered in linear approximation using the parameter dn_r/dT (Table 4.3).

4.3 Boundary Conditions

The behavior of the electromagnetic fields $\vec{E}$ and $\vec{H}$ at material boundaries is of great importance in the treatment of optoelectronic devices, which often feature different semiconductor layers. Let us consider a boundary between material 1 and material 2 with the electromagnetic fields at both sides of the boundary being represented by $\vec{E}_1, \vec{H}_1$ and $\vec{E}_2, \vec{H}_2$. The direction normal to the boundary, from medium 2 to medium 1, is given by the unit normal vector $\vec{\nu}$ (Fig. 4.6). By integrating the first two Maxwell equations over a small area across the boundary, the following boundary conditions are obtained for the tangential field components $E_\parallel$ and $H_\parallel$ parallel to the interface

$$\vec{\nu} \times (\vec{E}_1 - \vec{E}_2) = E_{\parallel,1} - E_{\parallel,2} = 0 \tag{4.35}$$

$$\vec{\nu} \times (\vec{H}_1 - \vec{H}_2) = H_{\parallel,1} - H_{\parallel,2} = \vec{j}_s, \tag{4.36}$$

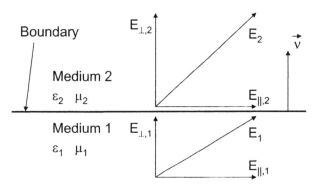

Figure 4.6: Illustration of the boundary conditions for the electric field $\vec{E}$.

where $\vec{j}_s$ (A/cm) is the surface current density of carriers moving along the interface. If $\vec{j}_s = 0$, both the tangential components $E_\|$ and $H_\|$ of the optical wave pass the semiconductor interface unchanged. Integrating the other two Maxwell equations over a small volume across the boundary, conditions for the normal field components $E_\perp$ and $H_\perp$ are obtained

$$\vec{v} \cdot (\varepsilon_1 \vec{F}_1 - \varepsilon_2 \vec{E}_2) = \varepsilon_1 E_{\perp,1} - \varepsilon_2 E_{\perp,2} = \varrho_s/\varepsilon_0 \tag{4.37}$$

$$\vec{v} \cdot (\mu_1 \vec{H}_1 - \mu_2 \vec{H}_2) = \mu_1 H_{\perp,1} - \mu_2 H_{\perp,2} = 0 \tag{4.38}$$

with ϱ_s (C/cm^2) representing the interface charge density. Thus, even when static charge densities are neglected, the normal component of the electrical field does change at material interfaces due to the difference in dielectric constants. The normal component $H_\perp$ of the magnetic field does not change at semiconductor boundaries if $\mu_1 = \mu_2 = 1$. However, at semiconductor–metal boundaries, interface charges are relevant. Optical waves disappear inside a perfect conductor due to screening by free electrons at the interface. This can be accounted for by using an appropriate imaginary part of the metal's dielectric constant.

4.4 Plane Waves

Optical waves in free space and in isotropic media are often approximated as plane waves for which $\vec{E}$ and $\vec{H}$ point in a constant direction and lie in a plane normal to the direction of the propagation given by the wave vector $\vec{k}$. Such a wave is also called transverse electromagnetic (TEM) wave. The fields are given as

$$\vec{E}(\vec{r}, t) = \vec{E}_0 \exp\left[i(\omega t - \vec{k}\vec{r})\right] \tag{4.39}$$

$$\vec{H}(\vec{r}, t) = \vec{H}_0 \exp\left[i(\omega t - \vec{k}\vec{r})\right], \tag{4.40}$$

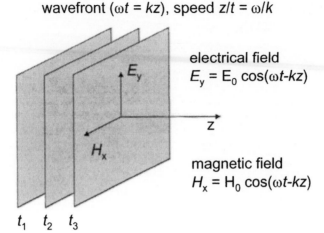

wavefront ($\omega t = kz$), speed $z/t = \omega/k$

electrical field
$E_y = E_0 \cos(\omega t - kz)$

magnetic field
$H_x = H_0 \cos(\omega t - kz)$

Figure 4.7: Illustration of a transverse electromagnetic (TEM) plane wave traveling in the z direction.

where $\vec{E}_0$ and $\vec{H}_0$ are constant vectors (Fig. 4.7). The plane of constant phase is defined by $\omega t - \vec{k}\vec{r} = \text{constant}$, and it travels with the phase velocity $c = \omega/k$. The wave number $k = |\vec{k}|$ depends on the refractive index $n_r = \sqrt{\varepsilon_{opt}}$ of the medium as

$$k = \sqrt{k_x^2 + k_y^2 + k_z^2} = \frac{2\pi}{\lambda} = n_r k_0 = n_r \frac{2\pi}{\lambda_0}, \qquad (4.41)$$

with the free-space values k_0 and λ_0. The phase velocity becomes $c = c_0/n_r$ with the free-space light velocity $c_0 = 1/\sqrt{\varepsilon_0\mu_0}$. Through Maxwell's equations, the two field amplitudes are related by

$$\vec{k} \times \vec{E}_0 = \omega\mu_0\vec{H}_0 \qquad (4.42)$$

$$\vec{k} \times \vec{H}_0 = -\omega\varepsilon\varepsilon_0\vec{E}_0 \qquad (4.43)$$

in addition to the divergence equations

$$\vec{k} \cdot \vec{E}_0 = 0 \qquad (4.44)$$

$$\vec{k} \cdot \vec{H}_0 = 0, \qquad (4.45)$$

which show that all three vectors are perpendicular to each other and

$$|\vec{H}_0| = \sqrt{\varepsilon\varepsilon_0/\mu_0}|\vec{E}_0|. \qquad (4.46)$$

The flow of electromagnetic energy is given by the Poynting vector

$$\vec{S} = \vec{E} \times \vec{H}, \tag{4.47}$$

whose time average gives the intensity (W/cm^2) of the optical wave as

$$I_{\text{opt}} = \sqrt{\frac{\varepsilon_{\text{opt}}\varepsilon_0}{4\mu_0}} |\vec{E}_0|^2. \tag{4.48}$$

Dividing by the photon energy gives the photon flux density (number of photons per area and per second)

$$\Phi_{\text{ph}} = I_{\text{opt}}/\hbar\omega. \tag{4.49}$$

The time-averaged local optical energy density (W/cm^3) is given by

$$W_{\text{opt}} = \varepsilon_{\text{opt}}\varepsilon_0 |\vec{E}_0|^2, \tag{4.50}$$

assuming an equal amount of energy in the electrical and magnetic components.

4.5 Plane Waves at Interfaces

When plane waves encounter a material boundary, the tangential and normal components of $\vec{E}_0$ and $\vec{H}_0$ are affected quite differently, as outlined in Section 4.3. Two specific field polarizations of TEM waves are usually considered, which are typical for multilayer waveguides. In transverse electric (TE) polarization, the electrical field is parallel to the interface ($\vec{E}_0^{\text{TE}} = E_y$ in Fig. 4.8). In transverse magnetic (TM) polarization, the magnetic field is parallel to the interface ($\vec{H}_0^{\text{TM}} = H_y$). In both cases, the angle of reflection ϑ_r is equal to the angle of incidence ϑ_i and the angle of transmission (angle of refraction) ϑ_t is given by Snell's law

$$n_{r1} \sin \vartheta_i = n_{r2} \sin \vartheta_t. \tag{4.51}$$

The boundary conditions lead to the following relations between the electric field strengths E_r, E_t, and E_i of reflected, transmitted, and incident waves, respectively. For TE waves, Eq. (4.35) gives

$$r_{12}^{\text{TE}} = \frac{E_r^{\text{TE}}}{E_i^{\text{TE}}} = \frac{n_{r1} \cos \vartheta_i - n_{r2} \cos \vartheta_t}{n_{r1} \cos \vartheta_i + n_{r2} \cos \vartheta_t} \tag{4.52}$$

$$t_{12}^{\text{TE}} = \frac{E_t^{\text{TE}}}{E_i^{\text{TE}}} = \frac{2n_{r1} \cos \vartheta_i}{n_{r1} \cos \vartheta_i + n_{r2} \cos \vartheta_t} \tag{4.53}$$

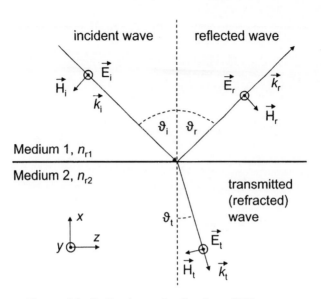

Figure 4.8: Reflection and refraction of TE waves.

with

$$\cos \vartheta_t = \sqrt{1 - n_{r1}^2 \sin^2 \vartheta_i / n_{r2}^2}. \tag{4.54}$$

The square of the field reflectivity coefficient r_{12} gives the power reflectivity (reflectance)

$$R = \left| \frac{E_r}{E_i} \right|^2 \tag{4.55}$$

and it is plotted in Fig. 4.9 for a GaAs–air interface. The important case of total reflection arises for $n_{r1} > n_{r2}$ when the angle of incidence is larger than a critical angle ϑ_c with

$$\sin \vartheta_c = n_{r2}/n_{r1}. \tag{4.56}$$

Equation (4.54) becomes imaginary in that case, which leads to an exponential decay of the transmitted field

$$E_t(x) = E_t(0) \exp \left[\frac{x}{d_p} \right] \tag{4.57}$$

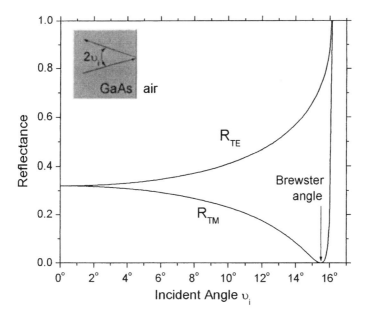

Figure 4.9: Power reflection coefficient (reflectance) for TE and TM waves as a function of the angle of incidence for $n_{r1} = 3.6$ (GaAs) and $n_{r2} = 1$ (air). The critical angle is 16.1°.

(cf. Fig. 4.8) with the penetration depth

$$d_p = \frac{\lambda_0}{2\pi n_{r2}\sqrt{n_{r1}^2 \sin^2 \vartheta_i - n_{r2}^2}} \qquad (4.58)$$

and the so-called Goos–Hänchen phase shift of the reflected field

$$E_r^{TE} = E_i^{TE} \exp[i 2\varphi_r^{TE}] \qquad (4.59)$$

with

$$\tan \varphi_r^{TE} = \frac{\sqrt{n_{r1}^2 \sin^2 \vartheta_i - n_{r2}^2}}{n_{r1} \cos \vartheta_i} \qquad (4.60)$$

(cf. Fig. 4.13). With the angle of incidence varying from ϑ_c to 90°, the phase shift $2\varphi_r^{TE}$ changes monotonically from 0 to π. Total reflection is the basis of waveguiding (see Section 4.8).

For TM waves, the magnetic field is parallel to the interface ($\vec{H}_0^{TM} = H_y$), and the electrical field lies in the plane of incident (normal to $\vec{k}$). The boundary conditions lead to

$$r_{12}^{TM} = \frac{E_r^{TM}}{E_i^{TM}} = \frac{n_{r2}\cos\vartheta_i - n_{r1}\cos\vartheta_t}{n_{r2}\cos\vartheta_i + n_{r1}\cos\vartheta_t} \tag{4.61}$$

$$t_{12}^{TM} = \frac{E_t^{TM}}{E_i^{TM}} = \frac{2n_{r1}\cos\vartheta_i}{n_{r2}\cos\vartheta_i + n_{r1}\cos\vartheta_t}. \tag{4.62}$$

The fundamental equations (4.52), (4.53), (4.61), and (4.62) are also referred to as Fresnel equations. The reflectance $|r_{12}|^2$ is plotted in Fig. 4.9 and it is lower for TM waves than for TE waves. It becomes zero for TM waves at the Brewster angle ϑ_B defined by

$$\tan\vartheta_B = n_{r2}/n_{r1} \tag{4.63}$$

Total reflection also occurs for TM waves with $\vartheta_i > \vartheta_c$. The penetration depth d_p of the transmitted wave is the same as for TE waves. The phase shift of the reflected wave $2\varphi_r^{TM}$ is now given by

$$\tan\varphi_r^{TM} = \frac{n_{r1}^2}{n_{r2}^2}\frac{\sqrt{n_{r1}^2\sin^2\vartheta_i - n_{r2}^2}}{n_{r1}\cos\vartheta_i}, \tag{4.64}$$

and it is larger than for TE waves. The power transmissivity (transmittance)

$$T = |t_{12}|^2\frac{n_{r2}\cos\vartheta_t}{n_{r1}\cos\vartheta_i} \tag{4.65}$$

is also different for TE and TM polarization. The correction factor after $|t_{12}|^2$ accounts for the different wave velocities in the normal direction. For normal incidence ($\vartheta_i = 0$), reflectance and transmittance of TE and TM mode are equal,

$$R = \left|\frac{\tilde{n}_1 - \tilde{n}_2}{\tilde{n}_1 + \tilde{n}_2}\right|^2 \tag{4.66}$$

$$T = \frac{n_{r2}}{n_{r1}}\left|\frac{2\tilde{n}_1}{\tilde{n}_1 + \tilde{n}_2}\right|^2, \tag{4.67}$$

which also applies to the more general case with complex refractive index $\tilde{n} = n_r + ik_a$ (see Section 4.2). The extinction coefficient k_a is usually very small in semiconductor waveguides. In contrast, metals have a very large k_a and strongly absorb optical waves. For that reason, Eq. (4.66) results in R ≈ 1 at metal interfaces.

Plane wave theory is used, for example, to track multiple reflections of light inside light-emitting diodes and solar cells (ray tracing).

4.6 Multilayer Structures

With several parallel interfaces, multiple reflections occur between the interfaces, and an infinite series of reflected and transmitted waves needs to be considered. This type of situation is often treated by the transfer matrix method, which shall be outlined in the following using the simple example of two parallel interfaces and normal incidence (Fig 4.10). At the first interface, the electric field on both sides can be represented by two counterpropagating waves

$$E_1 = E_1^- \exp[-ik_{1x}x] + E_1^+ \exp[+ik_{1x}x] \tag{4.68}$$

$$E_2 = E_2^- \exp[-ik_{2x}x] + E_2^+ \exp[+ik_{2x}x]. \tag{4.69}$$

All four components are connected by the boundary conditions for $\vec{E}$ and $\vec{H}$ fields, leading to

$$\begin{bmatrix} E_1^- \\ E_1^+ \end{bmatrix} = \begin{bmatrix} T_{11} & T_{12} \\ T_{21} & T_{22} \end{bmatrix} \begin{bmatrix} E_2^- \\ E_2^+ \end{bmatrix} = \hat{T} \begin{bmatrix} E_2^- \\ E_2^+ \end{bmatrix}, \tag{4.70}$$

where the coefficients of the transfer matrix $\hat{T}$ are given by $T_{11} = T_{22} = t_{12}^{-1}$ and $T_{12} = T_{21} = r_{12}t_{12}^{-1}$. The reflection coefficient r_{12} and the transmission coefficient t_{12} are defined in Section 4.5. The subsequent transmission through layer 2 of thickness d can also be described by a transfer matrix equation like Eq. (4.70) with $T_{11} = \exp[ik_{2x}d]$, $T_{22} = \exp[-ik_{2x}d]$, and $T_{12} = T_{21} = 0$. This is followed by another reflection. The overall transfer matrix of the layer is

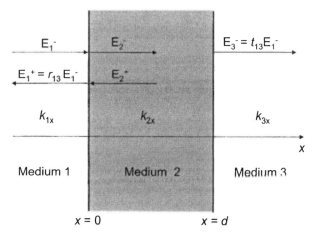

Figure 4.10: Normal reflection and transmission at two parallel interfaces.

obtained by sequential multiplication of all individual transfer matrices. This gives two equations for the two unknown field components E_1^+ and E_3^- of reflected and transmitted wave, respectively ($E_3^+ = 0$). The slab reflection and transmission coefficients become, respectively,

$$r_{13} = \frac{r_{12} + r_{23}\exp[-i2k_{2x}d]}{1 + r_{12}r_{23}\exp[-i2k_{2x}d]} \qquad (4.71)$$

$$t_{13} = \frac{t_{12}t_{23}\exp[-ik_{2x}d]}{1 + r_{12}r_{23}\exp[-i2k_{2x}d]}. \qquad (4.72)$$

These equations also hold for nonnormal incidence with the reflectance and transmittance given by

$$R = |r_{13}|^2 \qquad (4.73)$$

$$T = |t_{13}|^2 \frac{n_{r3}\cos\vartheta_3}{n_{r1}\cos\vartheta_1}, \qquad (4.74)$$

however, the coefficients r_{ij} and t_{ij} then have different values for TE and TM polarization (see Section 4.5). In general, the relations

$$r_{ij} = -r_{ji} \qquad (4.75)$$

$$t_{ij} = t_{ji} \qquad (4.76)$$

$$r_{ij}^2 + t_{ij}^2 = 1 \qquad (4.77)$$

are valid at any interface. If layer 2 is absorbing,

$$k_{2x} = \frac{2\pi}{\lambda_0}(n_{r2} + ik_{a2})\cos\vartheta_2 \qquad (4.78)$$

with the extinction coefficient k_{a2} and the absorption coefficient $\alpha_{o2} = 4\pi k_{a2}/\lambda$. The power absorbance A is given by optical energy conservation:

$$A = 1 - R - T. \qquad (4.79)$$

The transfer matrix method can easily be expanded to more than two interfaces [20, 122].

4.7 Helmholtz Wave Equations

The mathematical treatment of optical waveguides is based on the Helmholtz equations, which can be directly derived from the Maxwell equations

(4.6)–(4.9) as

$$\nabla^2 \vec{E} + k_0^2 \varepsilon \vec{E} = -\nabla \left[\vec{E} \cdot \frac{\nabla \varepsilon}{\varepsilon} \right] \tag{4.80}$$

$$\nabla^2 \vec{H} + k_0^2 \varepsilon \vec{H} = -\frac{\nabla \varepsilon}{\varepsilon} \times \nabla \times \vec{H}, \tag{4.81}$$

where $\varepsilon(\vec{r})$ is a function of the local position. In most semiconductor devices, $\varepsilon(\vec{r})$ is almost constant within each epitaxial layer, and it exhibits steplike changes only at layer boundaries. Thus, within each layer,

$$\nabla^2 \vec{E} + k_0^2 \varepsilon \vec{E} = 0 \tag{4.82}$$

$$\nabla^2 \vec{H} + k_0^2 \varepsilon \vec{H} = 0. \tag{4.83}$$

These homogeneous Helmholtz equations are typically solved within each homogeneous region and the solutions are matched at region interfaces, considering appropriate boundary conditions. Free-space solutions of the Helmholtz equation are plane waves (Section 4.4), Gaussian beams (Section 4.12), and spherical waves (Section 4.13). Each of these solutions applies to different observation ranges (Fig. 4.11).

For weak guiding, if the differentiation of the two polarization directions is not important, the vector Helmholtz equations (4.82) and (4.83) can be transferred into the scalar Helmholtz equation

$$\nabla^2 \Psi + k_0^2 \varepsilon \Psi = 0, \tag{4.84}$$

where $\Psi(x, y, z)$ represents the magnitude of the electrical or the magnetic field. If the propagation direction $\vec{k}$ is close to the z axis, the transformation

$$\Psi(x, y, z) = \Psi_T(x, y, z) \exp[-ikz] \tag{4.85}$$

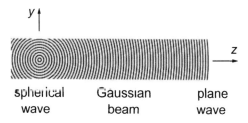

spherical   Gaussian   plane
wave   beam   wave

Figure 4.11: Validity range of spherical, paraboloidal (Gaussian beam), and plane waves in free space.

can be introduced, assuming that $\Psi_T(x, y, z)$ varies slowly in the z direction (paraxial approximation). $\Psi_T(x, y, z)$ is called the complex envelope of $\Psi(x, y, z)$, it may represent curved wavefronts as in Gaussian beams (Section 4.12). Neglecting higher order terms $\partial^2 \Psi_T / \partial z^2$ and $k^2 \Psi_T$, the paraxial Helmholtz equation becomes

$$\frac{\partial^2 \Psi_T}{\partial x^2} + \frac{\partial^2 \Psi_T}{\partial y^2} - i2k \frac{\partial \Psi_T}{\partial z} = 0. \qquad (4.86)$$

If the field $\Psi_T(x, y, z)$ is known at position z, the field at the next position $z + dz$ can be calculated from this equation (beam propagation method BPM [123]). Analytical solutions to the paraxial wave equation are discussed in Section 4.12.

Finally, with the plane wave approximation $\Psi = \Phi(x, y) \exp[i\beta z]$, we obtain the reduced scalar Helmholtz equation

$$\frac{\partial^2 \Phi}{\partial x^2} + \frac{\partial^2 \Phi}{\partial y^2} + (k^2 - \beta^2)\Phi = 0, \qquad (4.87)$$

where $\Phi(x, y)$ represents any transverse field component. Solving the scalar Helmholtz equation is in general much easier and faster than solving the vector Helmholtz equations. However, their validity in a given device as well as their boundary conditions need to be observed carefully. Reviews of the various numerical methods to solve Helmholtz's equations for optical waveguides are given in, e.g., [123, 124, 125]. In the following sections, we use simple examples to provide a general understanding of the resulting optical fields.

4.8   Symmetric Planar Waveguides

Symmetric planar waveguides are formed by a thin layer (refractive index n_{r1}) embedded in a cladding material with a slightly lower refractive index, n_{r2}. A plane optical wave traveling within the waveguide layer experiences total reflection if $\vec{k}$ has an angle θ to the waveguide plane that is less than the critical angle θ_c defined by

$$\cos \theta_c = \sin \vartheta_c = \frac{n_{r2}}{n_{r1}}. \qquad (4.88)$$

The bounce angle θ is complementary to the angle of incidence ϑ used in other sections. With $\theta < \theta_c$, the wave is considered guided. We will first employ the simplified ray picture in Fig. 4.12 to introduce some general features of guided waves. Only discrete bounce angles θ_m (discrete modes) are allowed for guided waves due to the self-consistency condition that a wave needs to

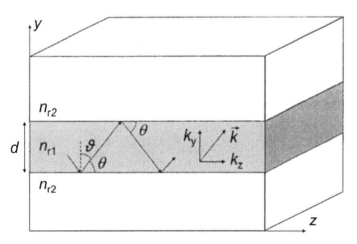

Figure 4.12: Simple representation of light beams in planar waveguides (θ, bounce angle; ϑ, angle of incidence; $\vec{k}$, wave vector).

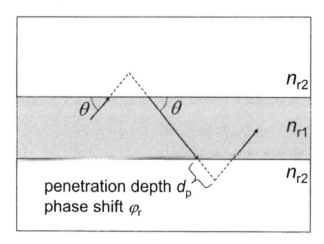

Figure 4.13: Goos–Hänchen phase shift $2\varphi_r$ with interface reflections. The penetration depth d_p can be on the order of the waveguide thickness.

reproduce itself after each round trip. The required phase shift for such constructive interference is

$$2dk \sin \theta_m - 2\psi_r = 2dk_{y,m} - 2\varphi_r = 2\pi m \quad (m = 0, 1, 2, \ldots), \qquad (4.89)$$

including a phase shift $2\varphi_r$ that accompanies each internal reflection. This Goos–Hänchen phase shift was introduced in Section 4.5 and it is illustrated in Fig. 4.13. It depends on the bounce angle and on the polarization in such a

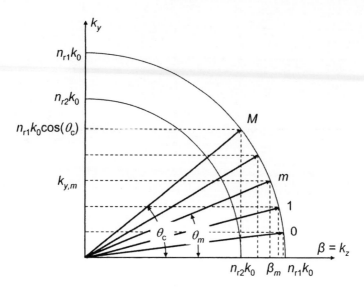

Figure 4.14: Modal wave numbers k_z and k_y in planar waveguides (n_{r1}, core index; n_{r2}, cladding index; m, mode number; θ_c, maximum bounce angle).

way that Eq. (4.89) needs to be solved numerically or graphically to obtain the modal angles θ_m [126]. The number of guided modes is limited by $\theta < \theta_c$, which results in

$$m < \frac{2d}{\lambda_0}\sqrt{n_{r1}^2 - n_{r2}^2}. \tag{4.90}$$

For any wavelength λ_0, the fundamental mode ($m = 0$) is always supported. However, θ_m rises with longer wavelength and comes closer to the cutoff angle θ_c, resulting in weaker guiding. The guided wave is composed of two distinct plane waves traveling at angles $\pm\theta$ with the z axis in the y–z plane (Fig. 4.12). The two wave vectors add up to the propagation constant of the guided wave

$$\beta_m = k_{z,m} = n_{r1}k_0 \cos\theta_m. \tag{4.91}$$

Since $\cos\theta_m$ lies between 1 and $\cos\theta_c = n_{r2}/n_{r1}$, β_m lies between $n_{r2}k_0$ and $n_{r1}k_0$ as illustrated in Fig. 4.14. The effective index of each mode is given by

$$n_{\text{eff},m} = \frac{\beta_m}{k_0} = n_{r1}\cos\theta_m \tag{4.92}$$

with the fundamental mode having the largest propagation constant and the largest effective index.

Two types of modal polarization are distinguished in planar waveguides:

- **TE modes** have a transversal electric field $E_0 = E_x$ (parallel to the waveguide plane and normal to the travel direction $\vec{k}$). The magnetic field $\vec{H}$ is also perpendicular to $\vec{k}$ and has components in the y and z directions.

- **TM modes** have a transversal magnetic field $H_0 = H_x$ (parallel to the waveguide plane and normal to the travel direction $\vec{k}$). The electrical field $\vec{E}$ is also perpendicular to $\vec{k}$ and has components in the y and z directions.

Let us now evaluate the TE field distribution $E_x(\vec{r})$ by solving the Helmholtz equation.[3] From the discussion above, $E_x(\vec{r})$ can be replaced by $E_x(y)\exp[i\beta z]$, and the Helmholtz wave equation simplifies to

$$\left[\frac{\partial^2}{\partial y^2} + (k^2 - \beta^2)\right]E_x(y) = 0, \tag{4.93}$$

which has different solutions in each region. Inside the waveguide,

$$k^2 - \beta_m^2 = k_{y,m}^2 = n_{r1}^2 k_0^2 \sin\theta_m \tag{4.94}$$

is a positive number, leading to a wavelike solution

$$E_{x,m}(y) \propto \cos(k_{y,m}y) \quad \text{for } m = 0, 2, 4, \ldots \tag{4.95}$$

$$E_{x,m}(y) \propto \sin(k_{y,m}y) \quad \text{for } m = 1, 3, 5, \ldots . \tag{4.96}$$

Outside the waveguide,

$$k^2 - \beta_m^2 = n_{r2}^2 k_0^2 - \beta_m^2 = -\gamma_m^2 \tag{4.97}$$

is negative, leading to an exponential solution

$$E_{x,m}(y) \propto \exp(-\gamma_m y) \quad \text{for } y > d/2 \tag{4.98}$$

$$E_{x,m}(y) \propto \exp(\gamma_m y) \quad \text{for } y < -d/2 \tag{4.99}$$

with the decay rate γ_m. The transversal field E_x is continuous at the boundary, giving the solutions shown in Fig. 4.15.

Besides guided modes, waveguides also exhibit radiative (leaky) modes, which have smaller propagation constants $\beta < n_2 k_0$ and bounce angles $\theta > \theta_c$. These quasi modes do not experience total reflection and they drain energy from the waveguide; however, they can still propagate inside the waveguide for quite some

---

[3]TM waves can be treated in a similar fashion [20].

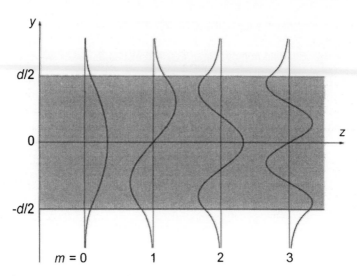

Figure 4.15: Planar waveguide solutions to the Helmholtz equation for the TE field $E_x(y)$.

distance as the reflectivity of the waveguide walls can still be relatively large [39, 127, 128]. Waveguide bends or facets may lead to an undesired transfer of optical power from guided modes to radiation modes (see Section 4.10). In waveguide simulations, the continuous spectrum of the radiation modes is often artificially discretized by employing homogenous Dirichlet or Neumann boundary conditions (electric wall or magnetic wall, respectively), sometimes supplemented by a so-called perfectly matched layer (PML) in order to avoid parasitic reflections from the boundaries [129].

4.9   Rectangular Waveguides

Two-dimensional waveguides confine the light in both transverse directions with thicknesses d_y and d_x. The principles and the modal structure in each direction are similar to planar waveguides. Vertical (y) and lateral (x) modes have independent indexes m and n, respectively. Instead of TE and TM polarization, two similar cases can be distinguished (Fig. 4.16):

- **HE_{mn} modes** are dominated by the lateral electric field E_x and the vertical magnetic field H_y, and

- **EH_{mn} modes** are dominated by the lateral magnetic field H_x and the vertical electric field E_y

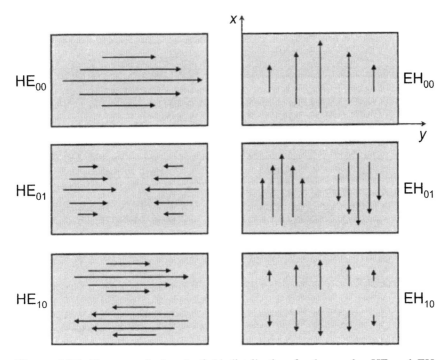

Figure 4.16: Transversal electric field distribution for low-order HE and EH modes in rectangular waveguides.

All other field components are also present, but they are small since the refractive index difference between core and cladding materials is small. Various numerical methods for calculating the modal fields have been developed. An approximate analytical method has been proposed by Marcatili [130].

Some examples of practical semiconductor waveguides are shown in Fig. 4.17. In cases like the ridge waveguide, the lateral confinement of the mode is achieved by changes in the vertical waveguide structure. To calculate guided modes in this kind of situation, the effective index method is frequently used. This method solves for the vertical modes in each lateral region and finds the effective indeces n_{eff}, which are different for each region. For a given vertical mode, the lateral modes are calculated based on the lateral steps between the different effective indeces. The effective index method is surprisingly accurate in determining the modal propagation constants; however, it tends to overestimate the modal confinement in the vertical direction and to underestimate the lateral confinement [131].

The optical field in different practical waveguide examples is further analyzed in Chapters 9 (laser diode), 10 (modulator), and 11 (amplifier photodetector).

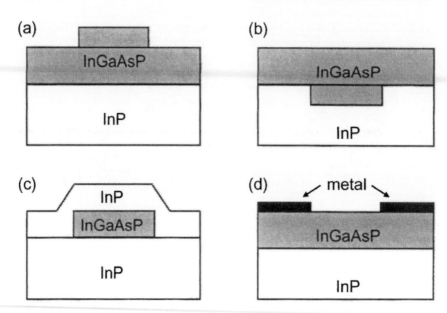

Figure 4.17: Examples of rectangular strip waveguides: (a) ridge waveguide, (b) inverted ridge waveguide, (c) buried waveguide, and (d) waveguide formed with metallic strips.

4.10 Facet Reflection of Waveguide Modes

The termination of the waveguide at a crystal-air facet results in partial reflection and partial transmission of the guided mode. In plane wave approximation with perpendicular angle of incidence, the facet reflectance is given by the Fresnel equation (Eq. (4.66)) using the effective modal index n_{eff} and the index of air. However, waveguide modes are not exactly plane waves and facet reflection of an incident mode transfers power into other guided modes as well as into radiative modes. The total internal field is composed of the incident guided mode (order m) and a superposition of all reflected modes (m'). It is here sufficient to consider the transverse (x, y) components of the electrical and magnetic field, respectively,

$$\vec{E}_t(x, y, z) = \vec{E}_{t,m}(x, y)e^{-i\beta_m z} + \sum_{m'} r_{mm'} \vec{E}_{t,m'}(x, y)e^{i\beta_{m'} z} \qquad (4.100)$$

$$\vec{H}_t(x, y, z) = \vec{H}_{t,m}(x, y)e^{-i\beta_m z} - \sum_{m'} r_{mm'} \vec{H}_{t,m'}(x, y)e^{i\beta_{m'} z}. \qquad (4.101)$$

The sum represents all reflected modes including radiation modes [127]. The normalized power reflection coefficient of mode m is given by

$$R_m = |r_{mm}|^2. \qquad (4.102)$$

Internal mode conversion coefficients can be obtained from the the nondiagonal elements of the matrix $r_{mm'}$. In addition, the transmitted field outside the waveguide must be known to establish proper boundary conditions (cf. Section 4.13). A rigorous treatment of this reflection process is rather complicated [132, 133, 134]. Figure 4.18 plots approximate solutions for the fundamental TE and TM modes of a symmetric waveguide [135] with

$$R_{TM}R_{TE} \approx R_0^2 = \left[\frac{n_{\text{eff}} + 1}{n_{\text{eff}} - 1}\right]^4. \qquad (4.103)$$

The Fresnel value R_0 varies slightly with the waveguide thickness due to the changing effective index ($n_1 > n_{\text{eff}} > n_2$); however, the actual modal reflectance exhibits significant deviations. The facet reflectivity for TE modes is usually higher

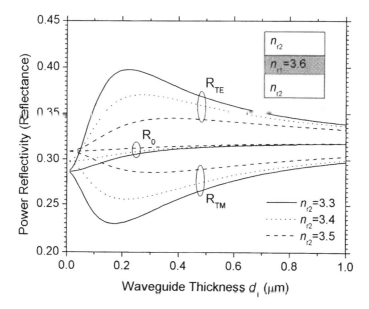

Figure 4.18: Facet reflectance of an AlGaAs-GaAs-AlGaAs waveguide for fundamental TE and TM modes, respectively, compared to the Fresnel reflectance R_0 (n_{r1}, n_{r2}, refractive indicees at 860 nm wavelength).

than for TM modes, resulting in a lower lasing threshold for TE modes. An optimum refractive index profile can be extracted, which gives maximum reflectance for the TE mode. The strongest variation of the refractive index usually occurs in vertical direction (y). Neglecting lateral variations, the following approximation was derived for arbitrary profiles $n_r(y)$ [136]

$$R_{TE} = \left[\frac{n_{eff} + 1 + \delta/2}{n_{eff} - 1 - \delta/2} \right]^2 , \qquad (4.104)$$

with the correction to Fresnel's formula

$$\delta = n_{eff}^2 - \frac{\int n_r^2(y)|E_x(y)|^2 \, dy}{\int |E_x(y)|^2 \, dy}. \qquad (4.105)$$

The TM reflection coefficient can be obtained from Eq. (4.103).

The common use of low- or high-reflection facet coatings further increases the difficulty of precise reflectivity computation [137]. Angled facets are utilized in semiconductor optical amplifiers, for instance, to obtain ultra low reflectivities [138].

4.11 Periodic Structures

In order to outline the optical properties of periodic structures, like distributed feedback (DFB) lasers, we assume that the refractive index varies in the direction of propagation as

$$n_r(z) = n_{eff} + \Delta n \cos(2\pi z/\Lambda) \qquad (4.106)$$

with the grating period (pitch) Λ and $\Delta n \ll n_{eff}$ (Fig. 4.19). Such a grating continuously reflects the incoming wave (distributed feedback). The reflected waves exhibit a constructive phase difference at the Bragg wavelength

$$\lambda = \frac{\lambda_0}{n_{eff}} = \frac{2\Lambda}{M} \qquad (4.107)$$

for the grating order $M = 1, 2, \ldots$ We assume a first order grating in the following, which gives the strongest reflected wave ($\Lambda \approx 235$ nm for 1st order gratings in InP at $\lambda_0 = 1.55\,\mu$m).[4] The Bragg propagation constant therefore is $\beta_B = \pi/\Lambda$. Neglecting variations in the transverse plane, the scalar Helmholtz equation becomes

$$\frac{\partial^2 \Phi(z)}{\partial z^2} + [n_r(z)k_0]^2 \Phi(z) \approx \frac{\partial^2 \Phi(z)}{\partial z^2} + [\beta^2 + 4\beta\kappa \cos(2\beta_B z)]\Phi(z) = 0 \qquad (4.108)$$

---

[4]Higher order gratings generate additional sideways radiation [139].

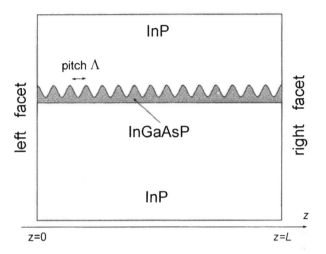

Figure 4.19: Distributed feedback structure (the facets typically exhibit antireflection coating).

with the propagation constant

$$\beta = \frac{2\pi n_{\text{eff}}}{\lambda_0} \qquad (4.109)$$

and the coupling coefficient

$$\kappa = \frac{\pi \Delta n}{\lambda_0} \qquad (4.110)$$

(the term containing Δn^2 was neglected). Let us now consider waves close to the Bragg condition $\beta = \beta_B$, with small deviations in the real part ($\delta \ll \beta_B$) and in the imaginary part (($g - \alpha_o$) $\ll \beta_B$)

$$\beta = \beta_B + \delta + i\frac{g - \alpha_o}{2} \qquad (4.111)$$

(the imaginary part represents gain or absorption). The total optical field $\Phi(z)$ consists of two counterpropagating waves that can be approximated as

$$\Phi(z) = \Phi^+(z)\exp[+i\beta_B z] + \Phi^-(z)\exp[-i\beta_B z]. \qquad (4.112)$$

Substituting $\Phi(z)$ into the wave equation (4.108) and comparing terms of equal phase, we obtain the coupled-mode equations

$$\frac{\partial \Phi^+}{\partial z} + \left(\frac{\alpha_0}{2} - i\delta\right)\Phi^+ = i\kappa\Phi^- \qquad (4.113)$$

$$-\frac{\partial \Phi^-}{\partial z} + \left(\frac{\alpha_0}{2} - i\delta\right)\Phi^- = i\kappa\Phi^+ \qquad (4.114)$$

(second derivatives are ignored for slowly varying envelopes Φ^+ and Φ^-). The two equations are coupled through the coupling coefficient κ, which depends on the shape of the grating. For other than cosine variations of the refractive index, the function $n_r(z)$ can be represented by a Fourier expansion. Taking a rectangular refractive index profile, for example, the first two Fourier terms are

$$n_r(z) = n_{\text{eff}} + \frac{4}{\pi}\Delta n \cos(2\beta_B z) + \dots, \qquad (4.115)$$

leading to the coupling constant $\kappa = 4\Delta n/\lambda_0$. For gratings that cover only part of the transverse mode extention, κ is weighted by a confinement factor. For more details on distributed feedback lasers, the reader is referred to specialized texts [139, 140].

4.12 Gaussian Beams

Gaussian beams are solutions to the Helmholtz equation in paraxial approximation (Section 4.7) given by the function

$$\Psi(x, y, z) = A_0 \frac{W_0}{W(z)} \exp\left[-\frac{x^2 + y^2}{W^2(z)} - i\left(kz + k\frac{x^2 + y^2}{2R(z)} + \zeta(z)\right)\right] \quad (4.116)$$

with

$$\text{the beam radius } W(z) = W_0\sqrt{1 + \left(\frac{z}{z_0}\right)^2}, \qquad (4.117)$$

$$\text{the waist radius } W_0 = \sqrt{\lambda z_0/\pi}, \qquad (4.118)$$

$$\text{the wavefront curvature radius } R(z) = z\left[1 + \left(\frac{z_0}{z}\right)^2\right], \qquad (4.119)$$

$$\text{and the phase retardation } \zeta(z) = \arctan(z/z_0). \qquad (4.120)$$

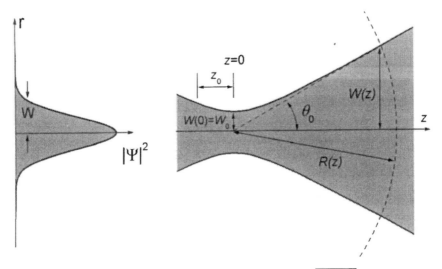

Figure 4.20: Gaussian beam intensity $|\Psi(r)|^2$ (left, $r = \sqrt{x^2 + y^2}$) and Gaussian beam radius $W(z)$ (right).

Figure 4.20 illustrates the Gaussian beam. At $z = 0$, the beam radius $W(z)$ is minimum (waist W_0) and the wavefront is planar (curvature radius $R(0) = \infty$). The beam waist diameter $2W_0$ is called spot size. As the beam travels in z direction, the convex wavefront first has a very large radius of curvature (i.e., it is almost planar), and it later approaches the curvature radius of a spherical wave ($R(z) = z$ for large z). The largest curvature occurs at $z = z_0$ with the smallest radius $R(z_0) = 2z_0$. The parameter z_0 is called Rayleigh range. For large z, the beam radius increases almost linear with z as the beam approaches the constant divergence angle

$$\theta_0 = \frac{\lambda}{\pi W_0}. \tag{4.121}$$

The beam divergence is stronger the smaller the beam waist is. On the beam axis, the phase $kz - \zeta(z)$ is that of a plane wave kz corrected by the phase retardation $\zeta(z)$, which ranges from $-\pi/2$ at $z = -\infty$ to $+\pi/2$ at $z = \infty$ with $\zeta(\pm z_0) = \pm\pi/4$. The intensity $I_{opt} = |\Psi|^2$ becomes

$$I_{opt}(x, y, z) = |A_0|^2 \left[\frac{W_0}{W(z)}\right]^2 \exp\left[-\frac{2(x^2 + y^2)}{W^2(z)}\right] \tag{4.122}$$

with the maximum value $|A_0|^2$ at $z = 0$ and half the peak value at $z = \pm z_0$. Integrating the intensity delivers the total beam power

$$P_{\text{opt}} = \frac{\pi}{2}|A_0|^2 W_0^2. \qquad (4.123)$$

About 86% of the power is contained within the beam radius $W(z)$, which marks $1/e^2$ of the axis intensity. The parameters A_0 and z_0 are obtained from boundary conditions.

The Gaussian beam approximation is often used to describe light beam transmission through small optical components for which ray optics fails. Upon transmission through circularly symmetric components aligned with the beam axis, the light beam remains a Gaussian beam as long as its paraxial nature is maintained. The transmission transfers the incoming wavefront curvature R_1 and beam radius W_1 into new outgoing values R_2 and W_2, respectively. Summarizing both parameters in a complex value

$$\frac{1}{q(z)} = \frac{1}{R(z)} - i\frac{\lambda}{\pi W^2(z)}, \qquad (4.124)$$

the transmission through an optical component can be described by using ABCD matrices (Fig. 4.21)

$$q_2 = \frac{Aq_1 + B}{Cq_1 + D}. \qquad (4.125)$$

For rectangular waveguides or other noncircular situations, the x and y directions can be separated in Eq. (4.116) and treated independently, giving an elliptical beam shape. This way, the astigmatism of laser diodes can be considered; i.e., different waist positions on the z axis for the x and y direction of the beam, respectively [141]. Higher order transverse modes are represented by Hermite–Gaussian functions as shown schematically in Fig. 4.22. More details on Gaussian beams can be found, for example, in [126, 142, 143].

4.13   Far Field

Laser diodes and similar devices emit an optical wave from the end of an internal waveguide into free space. In general, the larger the radiating aperture, the smaller the far field divergence. As shown in Fig. 4.23, the divergence along the axis perpendicular to the semiconductor layers (so-called fast axis) is usually much stronger than that along the axis parallel to the layers (slow axis). In some cases, the beam is astigmatic; i.e., the beam waists for both directions are located at different positions z [141]. If the ratio between the beam waists and the far field

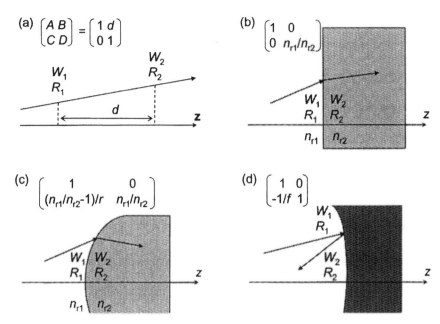

Figure 4.21: ABCD matrices for (a) beam translation, (b) flat refractive interface, (c) refractive interface with curvature radius r, and (d) reflection from a curved mirror with focal length f (n_{r1}, n_{r2}, refractive index; W_1, W_2, beam radius at interface; R_1, R_2, beam curvature radius at the interface).

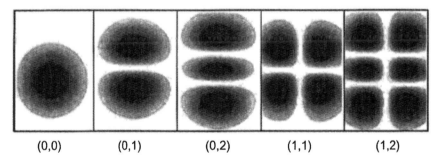

(0,0)   (0,1)   (0,2)   (1,1)   (1,2)

Figure 4.22: Transversal intensity distribution for Hermite–Gaussian beams of order (n, m).

divergences is nearly equal to the corresponding values of a Gaussian beam, the emitted field is called diffraction-limited. Especially the lateral far field of broad area lasers, strongly influenced by multi-mode behaviour and filamentation effects, is much wider than the diffraction limit.

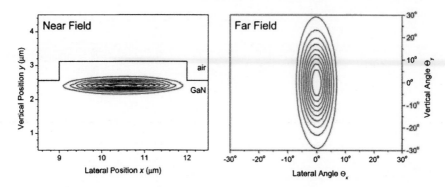

Figure 4.23: Near field and far field of a GaN laser diode (TE mode, cf. Section 9.4). The angles Θ_x and Θ_y are projections of Θ onto the x–z and the y–z plane, respectively (cf. Fig. 4.24).

The emitted field is obtained by solving the homogeneous Helmholtz equation for the free-space region [127]. It is closely related to the internal mode reflection as the fields left and right of the facet need to satisfy the electromagnetic boundary conditions. Assuming the facet at $z = 0$, the electric field just outside the waveguide is given by $\vec{E}^{NF}(x, y, 0)$ (near field). It is convenient to express the near field by its Fourier transform

$$\vec{E}^{NF}_{FT}(k_x, k_y) = \frac{1}{2\pi} \int_{-\infty}^{+\infty} dx \int_{-\infty}^{+\infty} dy \vec{E}^{NF}(x, y, 0) \exp[ik_x x + ik_y y]. \quad (4.126)$$

Neglecting reflection into other modes, $\vec{E}^{NF}_{FT}$ corresponds to the Fourier transform of the internally guided mode (cf. Section 4.10). The far field is the radiation field far away from the facet and it is given in spherical coordinates (r, Θ, Φ) as shown in Fig. 4.24 (cf. Section B.1). For large values of r (typically $r > 5\lambda_0$), the following analytical approximation can be obtained for the far field [136]

$$\vec{E}^{FF}(r, \Phi, \Theta) = -\frac{i \exp[-ik_0 r]}{k_0 r} \cos \Theta \, \vec{E}^{NF}_{FT}(k_0 \sin \Theta \cos \Phi, k_0 \sin \Theta \sin \Phi).$$
$$(4.127)$$

In order to compensate for the approximation errors introduced by this approach, the prefactor $\cos \Theta$ is often substituted by $(\cos \Theta)^2$ or by similar expressions [144]. This substitution also considers that the apparent emission area changes with the observation angle (Huygens obliquity factor [139]). The far field intensity

$$I^{FF}(\Phi, \Theta) = |\vec{E}^{FF}(r_0, \Phi, \Theta)|^2 \quad (4.128)$$

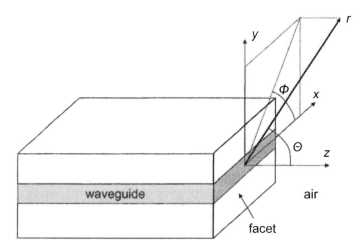

Figure 4.24: Far field coordinates (x, y, z) and (r, Θ, Φ).

is measured at a fixed distance $r = r_0$. For TE-like modes, defined by $E_y = 0$, we obtain from Eq. (4.127)

$$I_{\text{TE}}^{\text{FF}}(\Phi, \Theta) \propto \left[\cos^2 \Theta + \sin^2 \Theta \cos^2 \Phi \right] |E_{\text{FT},x}^{\text{NF}}(k_0 \sin \Theta \cos \Phi, k_0 \sin \Theta \sin \Phi)|^2.$$

(4.129)

Similarly, for TM-like modes ($E_x = 0$),

$$I_{\text{TM}}^{\text{FF}}(\Phi, \Theta) \propto \left[\cos^2 \Theta + \sin^2 \Theta \sin^2 \Phi \right] |E_{\text{FT},y}^{\text{NF}}(k_0 \sin \Theta \cos \Phi, k_0 \sin \Theta \sin \Phi)|^2.$$

(4.130)

The intensity profiles of near field and far field of a GaN-based laser diode are compared in Fig. 4.23 for the fundamental TE mode (cf. Chap. 9). Angled facets are discussed in [145].

Further Reading

- B. E. A. Saleh and M. C. Teich, *Fundamentals of Photonics*, Wiley, New York, 1991.

- A. W. Snyder and J. D. Love, *Optical Waveguide Theory*, Chapmann and Hall, London, 1991.

- K. J. Ebeling, *Integrated Optoelectronics*, Springer-Verlag, Berlin, 1993.

- H. P. Zappe, *Introduction to Semiconductor Integrated Optics*, Artech House, Boston, 1995.

- J. A. Kong, *Electromagnetic Wave Theory*, Wiley, New York, 1990.

- D. Marcuse, *Theory of Dielectric Optical Waveguides*, Academic Press, Boston, 1991.

- K. Okamoto, *Fundamentals of Optical Waveguides*, Academic Press, San Diego, 2000.

- K. Kawano and T. Kitoh, *Introduction to Optical Waveguide Analysis*, Wiley, New York, 2001.

Chapter 5

Photon Generation

In semiconductors, photons are typically generated by electron transitions from the conduction band to the valence band. This transition can happen spontaneously or it can be stimulated by another photon. Stimulated photon emission generates optical gain, which is key to laser diodes and semi-conductor optical amplifiers. This chapter mainly focuses on optical gain. Spontaneous emission of photons is key to light-emitting diodes, and it is covered at the end.

5.1 Optical Gain

Optical gain is defined as the growth ratio of light intensity (photon density) per unit length of light propagation. Like the optical absorption coefficient α_o, the optical gain coefficient g is given in inverse centimeters (1/cm). An incoming photon can either be absorbed or generate gain. Photon absorption causes the transition of an electron from a lower to a higher energy band, creating an electron–hole pair. Gain is generated by stimulated recombination of an existing electron–hole pair, creating a second photon. The second photon exhibits the same wavelength and the same phase as the first photon, doubling the amplitude of the monochromatic wave. Subsequent repetition of this process leads to strong light amplification. However, the competing process is the absorption of photons by the generation of new electron–hole pairs. Stimulated emission prevails when more electrons are present at the higher energy level (conduction band) than at the lower energy level (valence band). This inversion of the carrier population can be achieved at *pn*-junctions by providing conduction band electrons from the *n*-doped side and valence band holes from the *p*-doped side. At low injection current, band-to-band absorption still dominates and the optical gain g is below zero. At the transparency current, both processes are equally strong, the gain is zero, and the material is transparent. Even stronger current causes net amplification of light.

The optical gain is proportional to the probability that a given photon triggers an electron transition from a higher energy level j to a lower energy level i. The photon energy $h\nu$ must be equal to the transition energy $E_{ij} = E_j - E_i$. The quantum-mechanical calculation of this probability for semiconductors has been described in many publications (see, e.g., [10, 20]). To provide a more intuitive understanding, we skip most of the quantum mechanics here and evaluate the

121

simple gain function [146]

$$g_{ij}(h\nu) = \left(\frac{q^2 h}{2m_0^2 \varepsilon_0 n_r c}\right)\left(\frac{1}{h\nu}\right)|M(E_{ij})|^2 D_r(E_{ij})(f_j - f_i), \qquad (5.1)$$

for $h\nu = E_{ij}$. The main parameter in this equation is the transition matrix element $|M|^2$, which determines the transition strength between both electron levels (see below). $D_r(E)$ is the density of allowed transitions between the two bands. It is called the reduced density of electron states and, in the case of bulk semiconductors, is calculated as

$$D_r(E) = \left[\frac{1}{D_c} + \frac{1}{D_v}\right]^{-1} = \frac{1}{2\pi^2}\left(\frac{2m_r}{\hbar^2}\right)^{\frac{3}{2}}\sqrt{E - E_g} \qquad (5.2)$$

$(E > E_g)$ with the reduced effective mass

$$m_r = \left[\frac{1}{m_c} + \frac{1}{m_v}\right]^{-1}. \qquad (5.3)$$

For quantum wells of thickness d_z the reduced density of each subband is

$$D_r^{2D} = \frac{m_r}{\pi \hbar^2 d_z}. \qquad (5.4)$$

With higher energy, more and more subbands are added up in calculating the total gain (cf. Fig. 2.14). Allowed are only vertical transitions within the $E(\vec{k})$ diagram since the electron momentum as well as the electron spin must be the same in the higher and the lower states (k-selection rule). Additional selection rules apply in quantum wells, as illustrated in Fig. 5.1. Only transitions between subbands with the same quantum number m are allowed. Transitions between subbands with dissimilar quantum numbers are forbidden transitions. However, valence band mixing enables forbidden transitions at larger k vectors (see Section 5.1.1).

The Fermi functions f_i and f_j in Eq. (5.1) give the probability that the energy levels E_i and E_j, respectively, are occupied by electrons. Maximum gain is obtained with $f_j = 1$ at the upper level and $f_i = 0$ at the lower level. The opposite case gives maximum absorption. The Fermi factor $(f_j - f_i)$ results from $f_j(1 - f_i) - f_i(1 - f_j)$ accounting for stimulated emission and band-to-band absorption. The stimulated emission rate, for example, is proportional to the occupation of the higher level (f_j) as well as to the chance that the lower level is empty $(1 - f_i)$. Figure 5.2 shows schematically how the gain varies with photon energy. The lower energy limit of positive gain is set by the reduced density of states. Maximum gain is reached for $f_i - f_j = 1$. However, the Fermi

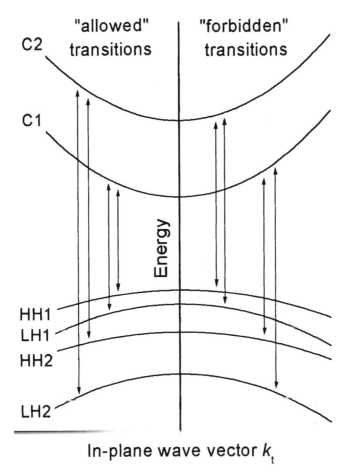

Figure 5.1: Allowed and forbidden transitions in a quantum well (C#, conduction band; HH#, heavy-hole band; LH#, light-hole band; #, subband number).

factor decays with higher energy, and it becomes zero when the energy equals the quasi-Fermi level distance ΔE_F. Band-to-band absorption dominates for higher photon energies (negative gain). Thus, the gain spectrum is generally limited to $E_g < h\nu < \Delta E_F$. Compared to the smooth gain spectrum in bulk material, quantum wells are expected to exhibit rather sharp features due to the constant density of states for each subband (Fig. 5.2). However, real quantum well gain spectra do not show such sharp features due to transition energy broadening (dephasing), which is discussed in Section 5.1.2.

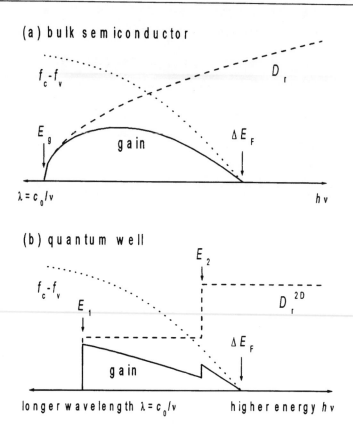

Figure 5.2: Schematic gain spectra for (a) bulk semiconductors and (b) quantum wells (cf. Eq. (5.1)).

5.1.1 Transition Matrix Element

The transition matrix element $|M|^2$ is a measure for the strength of stimulated electron transitions in a given material. This strength does not depend on the direction of the interband transition; it is the same for emission and absorption. However, the transition strength does depend on the angle between the electron wave vector $\vec{k}$ and the optical field vector $\vec{E}$. Any polarization direction of the optical field encounters a variety of electron $\vec{k}$ vectors that needs to be averaged at the given photon energy. For bulk zinc blende semiconductors, averaging over all possible $\vec{k}$ vectors results in an isotropic transition matrix element that is equal to the momentum matrix element introduced in Section 2.2.3,

$$|M|^2 = M_{\mathrm{b}}^2 = \frac{m_0}{6} E_{\mathrm{p}} = \frac{m_0}{6} \left(\frac{m_0}{m_{\mathrm{c}}} - 1 - 2F_{\mathrm{b}} \right) \frac{E_{\mathrm{g}}(E_{\mathrm{g}} + \Delta_0)}{E_{\mathrm{g}} + 2\Delta_0/3}, \tag{5.5}$$

Table 5.1: Energy Parameter E_p of the Bulk Momentum Matrix Element M_b, Correction Factor F_b in Eqs. (5.5) and (2.61) [13], and Longitudinal Optical Phonon Energy $\hbar\omega_{LO}$ [2, 89] as Used in the Asada Scattering Model (Section 5.1.2)

Parameter	E_p	F_b	$\hbar\omega_{LO}$
Unit	(eV)	—	(meV)
GaAs	28.8	−1.94	36.2
InP	20.7	−1.31	42.8
AlAs	21.1	−0.48	50.1
GaSb	27.0	−1.63	28.7
AlSb	18.7	−0.56	42.1
InAs	21.5	−2.9	30.2
GaP	31.4	−2.04	50.7
AlP	17.7	−0.65	61.9
InSb	23.3	−0.23	24.2

Note. Parameters for wurtzite compounds GaN, InN, and AlN are listed in Table 2.7.

for electrons close to the Γ point. F_b is a correction parameter for the electron effective mass (cf. Eq. (2.61)). Electron spin resonance techniques allow for fairly accurate measurements of the momentum matrix element (Table 5.1).

For quantum well structures, the transition matrix element is anisotropic and the gain depends on the optical polarization. Commonly, one distinguishes two polarization modes, in which either the electric field (TE mode) or the magnetic field (TM mode) lies within the quantum well xy-plane (transversal plane). The transition strengths are different for heavy (hh) and light holes (lh)

$$|M_{hh}^{TE}|^2 = \frac{3 + 3\cos^2(\theta_e)}{4} O_{ij} M_b^2 \tag{5.6}$$

$$|M_{lh}^{TE}|^2 = \frac{5 - 3\cos^2(\theta_e)}{4} O_{ij} M_b^2 \tag{5.7}$$

$$|M_{hh}^{TM}|^2 = \frac{3 - 3\cos^2(\theta_e)}{2} O_{ij} M_b^2 \tag{5.8}$$

$$|M_{lh}^{TM}|^2 = \frac{1 + 3\cos^2(\theta_e)}{2} O_{ij} M_b^2. \tag{5.9}$$

These equations consider the angle θ_e of the electron $\vec{k}$ vector with the k_z direction

$$k_z = |\vec{k}| \cos(\theta_e) \tag{5.10}$$

with $\cos(\theta_e) = 1$ at the Γ point of the quantum well subband. The matrix element depends on the photon energy through the quantum well dispersion functions $E_m(\vec{k})$. It is different for each subband m. The overlap integral O_{ij} of the two quantum well envelope wave functions can assume values between 0 and 1. At the Γ point, O_{ij} is nonzero only for subbands with the same quantum number m (allowed transitions). Away from the Γ point, O_{ij} is nonzero even for forbidden transitions (see Fig. 5.3). Thus, at higher photon energies, summation over all possible subband combinations must be included in the gain calculation for quantum wells.

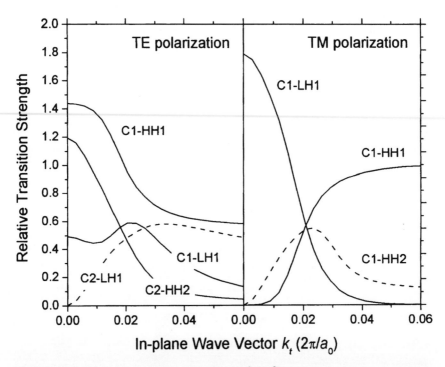

Figure 5.3: Relative transition strength $|M|^2/M_b^2$ for an unstrained 8-nm-thick GaAs/Al$_{0.15}$Ga$_{0.85}$As quantum well as a function of the in-plane wave vector for TE and TM polarization, respectively. Selected transitions are shown from the first quantum level (C1) and from the second quantum level (C2). Allowed transitions are plotted as solid lines; forbidden transitions as dashed lines. The transition energies are 1.468 eV (C1–HH1), 1.483 eV (C1–LH1), 1.497 eV (C1–HH2), 1.561 eV (C2–HH1), 1.576 eV (C2–LH1), and 1.590 eV (C2–HH2), respectively.

For both polarization directions, the relative transition strength $|M|^2/M_b^2$ is plotted in Fig. 5.3 as a function of the in-plane wave vector k_t. We here use the same GaAs/AlGaAs quantum well as in Fig. 2.15a, which supports two quantum levels in the conduction band (C1, C2). Heavy-hole transitions are favored under TE polarization while the light-hole transition is favored for TM polarization. In the TM case, the light-hole transition C1–LH1 has a larger strength than the heavy-hole transition C1–HH1 under TE polarization. Both of them are stronger than in bulk material. Thus, the use of quantum wells enhances the optical gain. Figure 5.3 also shows some of the forbidden transitions (dashed) whose transition strengths vanish at the Γ point.

The wurtzite crystal structure, as in GaN-based blue laser diodes, leads to different formulas for the transition matrix element. The bulk matrix element is not isotropic anymore and for a quantum well grown in the hexagonal c direction, the transition matrix elements for heavy (hh) and light holes (lh) are [59]

$$|M_{\text{hh}}^{\text{TE}}|^2 = \tfrac{3}{2} O_{ij} (M_{\text{b}}^{\text{TE}})^2 \tag{5.11}$$

$$|M_{\text{lh}}^{\text{TE}}|^2 = \tfrac{3}{2} \cos^2(\theta_e) O_{ij} (M_{\text{b}}^{\text{TE}})^2 \tag{5.12}$$

$$|M_{\text{ch}}^{\text{TE}}|^2 = 0 \tag{5.13}$$

$$|M_{\text{hh}}^{\text{TM}}|^2 = 0 \tag{5.14}$$

$$|M_{\text{lh}}^{\text{TM}}|^2 = \tfrac{3}{2} \sin^2(\theta_e) O_{ij} (M_{\text{b}}^{\text{TM}})^2 \tag{5.15}$$

$$|M_{\text{ch}}^{\text{TM}}|^2 = \tfrac{3}{2} O_{ij} (M_{\text{b}}^{\text{TM}})^2. \tag{5.16}$$

with anisotropic bulk momentum matrix elements [34]

$$\left(M_{\text{b}}^{\text{TM}}\right)^2 = \frac{m_0}{6} \left(\frac{m_0}{m_{\text{c}}^z} - 1\right) \frac{(E_g + \Delta_1 + \Delta_2)(E_g + 2\Delta_2) - 2\Delta_3^2}{E_g + 2\Delta_2} \tag{5.17}$$

$$\left(M_{\text{b}}^{\text{TE}}\right)^2 = \frac{m_0}{6} \left(\frac{m_0}{m_{\text{c}}^t} - 1\right) \frac{E_g[(E_g + \Delta_1 + \Delta_2)(E_g + 2\Delta_2) - 2\Delta_3^2]}{(E_g + \Delta_1 + \Delta_2)(E_g + \Delta_2) - \Delta_3^2}. \tag{5.18}$$

Note that the bulk electron mass is different in transversal (m_{c}^t) and in parallel directions (m_{c}^z) relative to the hexagonal c axis (cf. Table 2.7).

5.1.2 Transition Energy Broadening

Electrons and holes frequently interact with other carriers and with phonons, thereby changing their energy within the (sub)band. Such intraband scatter events happen about every 0.1 ps, much more often than band-to-band recombination events. Thus, scattering leads to an uncertainty of the electron energy, which can

be accounted for by introducing a symmetrical linewidth broadening function L
into the gain formula (Eq. (5.1))

$$g(h\nu) = \int dE_{ij} g_{ij}(E_{ij}) L(h\nu - E_{ij}). \qquad (5.19)$$

This convolution integral means that gain at the photon energy $h\nu$ can now receive
contributions from electron transitions with $E_{ij} \neq h\nu$, weighted by $L(h\nu - E_{ij})$.
In fact, positive gain is now possible even for photon energies slightly below the
band gap. Commonly, the Lorentzian line shape function,

$$L(h\nu - E_{ij}) = \frac{1}{\pi} \frac{\Gamma_s}{(h\nu - E_{ij})^2 + \Gamma_s^2}, \qquad (5.20)$$

is used with the half-width Γ_s (Fig. 5.4). This function is based on the assumption
that the occupation probability of an electron state decays proportionally to
$\exp(-t/\tau_s)$. The Fourier transformation of this exponential function into the energy
domain leads to Eq. (5.20). Γ_s is the average of the broadening in the conduction
and in the valence band. The full linewidth $2\Gamma_s$ is related to the average intraband

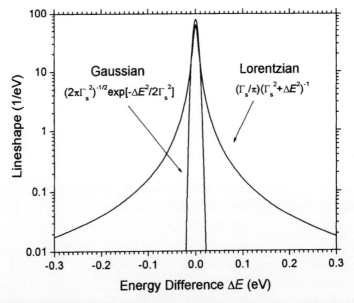

Figure 5.4: Line shape functions for a full linewidth of $2\Gamma_s = 0.01\,\mathrm{eV}$ ($\Delta E = h\nu - E_{ij}$, cf. Eq. (5.20)).

scattering time τ_s by

$$2\Gamma_s = 2\frac{\hbar}{\tau_s} = \frac{\hbar}{\tau_c} + \frac{\hbar}{\tau_v} = 2\sum_i \frac{\hbar}{\tau_{c,i}} + 2\sum_i \frac{\hbar}{\tau_{v,i}}$$

$$= \Gamma_c + \Gamma_v = 2\sum_i \Gamma_{c,i} + 2\sum_i \Gamma_{v,i}, \tag{5.21}$$

which includes scattering events in the conduction band (c) and valence band (v). For each band, linewidth contributions Γ_i from different scattering processes are adding up. With τ_s being typically on the order of 0.1 ps, the full linewidth is about $2\Gamma_s = 13$ meV. Gain spectra with different linewidth parameters are plotted in Fig. 5.5 for the same 8-nm GaAs/AlGaAs quantum well as in previous figures. Less broadening gives higher peak gain and better resolution of the quantum well subband levels. Note that the spectra are now plotted versus the photon wavelength $\lambda = c_0/\nu$ (cf. 5.2). Figure 5.6 plots gain spectra with different carrier

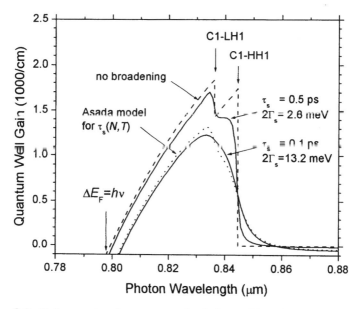

Figure 5.5: Gain spectra for an unstrained 8-nm-thick GaAs/Al$_{0.15}$Ga$_{0.85}$As quantum well with different Lorentzian linewidths $2\Gamma_s$ and scattering times τ_s for carrier concentrations $n = p = 5 \times 10^{18}$ cm^{-3} ($T = 300$ K). The dashed line gives the gain without broadening, which becomes zero when the separation ΔE_F of the quasi-Fermi levels equals the photon energy $h\nu$. The dotted line gives the result of the Asada model.

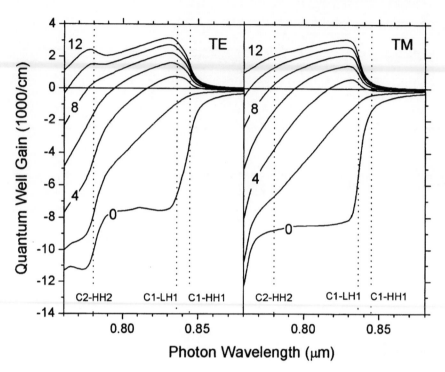

Figure 5.6: Gain spectra for an unstrained 8-nm-thick GaAs/Al$_{0.15}$Ga$_{0.85}$As quantum well with the carrier concentration as parameter (10^{18} cm^{-3}) ($\tau_s = 0.1$ ps, $T = 300$ K). The dotted lines indicate the relevant transition energies; band gap shrinking is neglected here.

concentrations. With vanishing carrier concentration, the quantum well exhibits strong band-to-band absorption and the absorption spectrum reflects the different quantum levels. Positive gain is obtained only with considerable carrier injection into the quantum well. At very high carrier concentrations, transitions between the second quantum levels start to contribute to the gain.

The linewidth is typically used as fit parameter to find agreement with experimental gain spectra. However, the long-wavelength side of measured spectra is not reproduced very well by using the Lorentzian line shape function. In fact, the low-energy tail of this function causes unphysical absorption below the band gap; i.e., for photon wavelengths above the C1–HH1 transition in Fig. 5.5. This artifact can be avoided by assuming a faster occupation decay rate, resulting in a narrower line shape, as represented by the Gaussian line shape in Fig. 5.4. A combination of both these line shape functions seems to be most reasonable [147].

In general, the scattering time τ_s reflects carrier collisions with other particles, and it should depend on carrier energy, carrier concentration, and temperature. Landsberg has suggested an energy-dependent half-width $\Gamma_s(E_{ij})$ that varies from a maximum value Γ_0 near the bandedge to 0 at the quasi-Fermi level separation ΔE_F [148, 149]

$$\frac{\Gamma_s(E_{ij})}{\Gamma_0} = 1 - 2.229 \frac{E_{ij}}{\Delta E_F} + 1.458 \left(\frac{E_{ij}}{\Delta E_F}\right)^2 - 0.229 \left(\frac{E_{ij}}{\Delta E_F}\right)^3 \quad (5.22)$$

for $E_g \leq E_{ij} \leq \Delta E_F$. This model has been originally developed for bulk material, and it is only valid for positive gain ($\Gamma_0 = 1.2\,\text{meV}$ for bulk GaAs). The Landsberg linewidth model has also been successfully employed to reproduce measured gain spectra of quantum wells where the broadening is much larger. For GaAs/AlGaAs wells, $2\Gamma_0 = 20\,\text{meV}$ ($\tau_s = 66\,\text{fs}$) was obtained, as well as $2\Gamma_0 = 30\,\text{meV}$ ($\tau_s = 43\,\text{fs}$) for InGaAs/InP quantum wells [150, 151].

However, the above models for the transition energy broadening are more phenomenological and a rigorous approach should consider the electron scattering processes in detail. Asada has developed a model for the Lorentzian line shape approximation that considers carrier–carrier scattering as well as carrier–phonon scattering within the conduction and valence bands [152, 153]. The model assumes parabolic bands, and it includes the screening of the Coulomb potential by other carriers. The optical dielectric constant ε_{opt} and the longitudinal optical (LO) phonon energy $\hbar\omega_{LO}$ are used as material parameters (see Tables 4.3 and 5.1). Results for our GaAs/AlGaAs quantum well are plotted in Fig. 5.7 as a function of carrier concentration and temperature, respectively. Broadening variations arise from changes in Fermi spreading and screening length. The largest contributions to the total broadening in our quantum well result from hole–hole and hole–LO phonon scattering within the valence band. Electron–hole scattering dominates in the conduction band. The total scattering time decreases with higher temperature and with lower carrier concentration. At room temperature and with $n = p = 5 \times 10^{18}\,\text{cm}^{-3}$, the scattering time is 87 fs. Thus, for our GaAs/AlGaAs quantum well, the Asada gain spectrum (dotted in Fig. 5.5) is close to the spectrum with a constant Lorentz scattering time of $\tau_s = 100\,\text{fs}$. More advanced models including asymmetric line shape functions are described in [153].

5.1.3 Strain Effects

The application of strain leads to a deformation of the band structure as outlined in Section 2.2. The thickness of the strained layer is typically only a few nanometers and it must be below the critical thickness to prevent lattice relaxation [6]. Stained quantum wells are widely utilized in optoelectronics as they allow for improvements of material properties like the optical gain [154]. Commonly, strain is

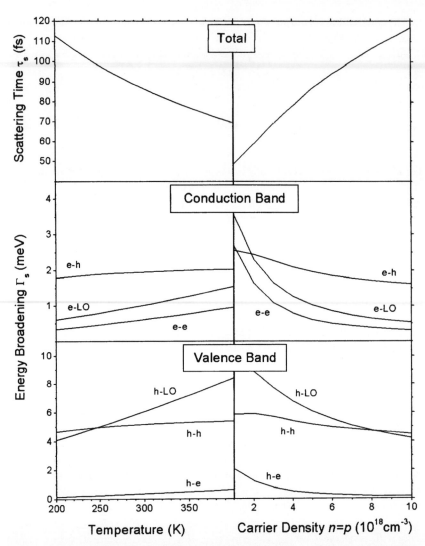

Figure 5.7: Results of the Asada scattering model for an unstrained 8-nm-thick GaAs/Al$_{0.15}$Ga$_{0.85}$As quantum well as a function of temperature T (left, $n = p = 5 \times 10^{18}\,\mathrm{cm}^{-3}$) and carrier concentration $n = p$ (right, $T = 300\,\mathrm{K}$). Top: average intraband scattering time. Middle: half-width of the energy broadening resulting from electron–electron (e–e), electron–hole (e–h), and electron–LO phonon scattering (e–LO) within the conduction band. Bottom: half width of the energy broadening resulting from hole–hole (h–h), hole–electron (h–e), and hole–LO phonon scattering (h–LO) within the valence band. Calculated with $\varepsilon_{\mathrm{opt}} = 10.9$ and $\hbar\omega_{\mathrm{LO}} = 36\,\mathrm{meV}$.

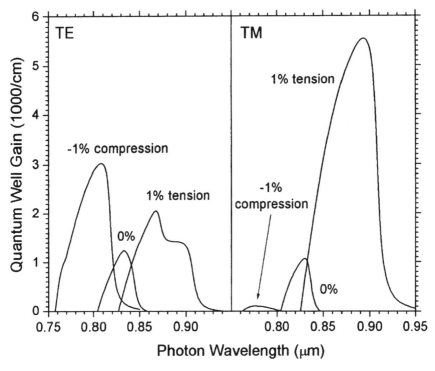

Figure 5.8: Gain spectra of our 8-nm-thick GaAs/Al$_{0.15}$Ga$_{0.85}$As quantum well at different GaAs strains for TE and TM polarization ($n = p = 5 \times 10^{18}$ cm^{-3}, $T = 300$ K).

generated by varying the materials composition so that the lattice constant slightly deviates from that of the substrate. However, to observe pure strain effects, we here apply strain to our GaAs quantum well without changing the composition. For the sake of simplicity, the AlGaAs barrier layers remain unstrained and the carrier concentration is kept constant. Figure 5.8 shows gain variations calculated for this theoretical example. The TM gain increases dramatically with tensile strain, whereas it almost disappears for compressive strain. The TE gain is improved in both cases. Even with our small amount of 1% strain, the transition energy changes significantly. In many practical applications, this wavelength shift needs to be compensated for by changing the layer composition.

To guide our understanding of these gain variations, the corresponding valence band structures and transition strengths are plotted in Fig. 5.9. The lower boundary gives the constant position of the AlGaAs valence band edge. Compression slightly lowers the HH band edge, and it moves the LH band edge below the barrier edge.

Thus, only heavy holes "see" a quantum well in this case. However, this band edge variation with compression seems to contradict the tendency given in Fig. 2.9, which suggests a deeper well. Indeed, at 1% compression, the model-solid theory predicts a quantum well that is 118 meV deep for heavy holes and 50 meV deep for light holes (cf. Section 2.5). This prediction is not in agreement with experimental results for the AlAs/GaAs valence band offset [155]. As an alternative approach, we here use the assumption of a constant band offset ratio $\Delta E_v/\Delta E_g = 0.35$ for small strain. Since the band gap E_g increases with compression, the band offset ΔE_v slightly shrinks as shown in Fig. 5.9. The elimination of the LH1 subband leaves more carriers for the C1–HH1 transition so that the TE peak gain is higher than in the unstrained case. Without light holes, the TM gain vanishes. With tensile strain, the light-hole band is on top (cf. Fig. 2.9). The lower band gap now gives a deeper quantum well for light holes. Due to the more shallow HH quantum well, the C1–LH1 transition is enhanced, which results in substantial TM gain improvement. However, this transition also contributes to TE gain, as well as the C1–HH1 transition. Both transition energies are visible in the TE gain spectrum

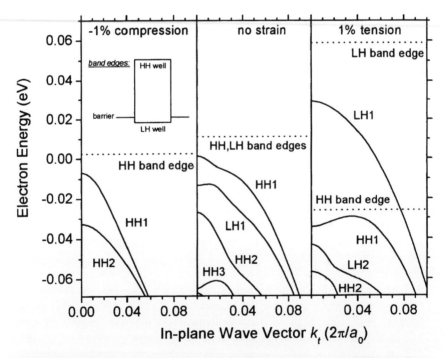

Figure 5.9: Valence band structure with compressive (left), no (middle), and tensile strain (right) for our 8-nm GaAs/Al$_{0.15}$Ga$_{0.85}$As quantum well. The horizontal lines indicate band edge positions.

of Fig. 5.8. As illustrated in this example, gain improvements in strained quantum wells mainly result from the increased separation of LH and HH subbands. Strain effects on device performance are further discussed in Chapters 9 and 10.

5.1.4 Many-Body Effects

Broadening effects due to scattering are only one example for the influence of carrier–carrier interaction on the gain. The screening of the electron Coulomb potential by surrounding carriers reduces the repulsion between valence and conduction band electrons and, among other factors, leads to a band gap reduction (cf. Section 2.1.3).

Another important many-body effect is the Coulomb enhancement of the gain due to mutual attraction of electrons and holes. For our GaAs quantum well, Coulomb enhancement effects on the gain are shown in Fig. 5.10 as

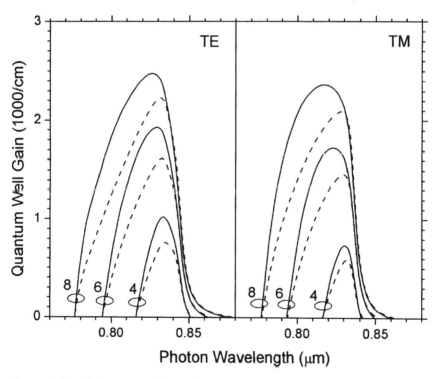

Figure 5.10: Gain spectra with (solid) and without (dashed) Coulomb enhancement, calculated for our 8-nm-thick GaAs/Al$_{0.15}$Ga$_{0.85}$As quantum well at TE and TM polarization with the carrier concentration given as a parameter (10^{18} cm^{-3}). The band gap reduction is neglected here.

calculated following the model in [156]. The blue-shift of the gain peak with rising carrier concentration is typically stronger than the red-shift from band gap renormalization.

A rigorous many-body theory of the gain including carrier collisions (Coulomb correlation) leads to a further enhancement and a blue-shift of the gain peak at high carrier concentrations. A thorough consideration of those many-body effects can be found elsewhere [10].

At low temperatures or low carrier concentrations, an electron–hole pair can form a hydrogen-like bond. This so-called exciton gives sharp absorption peaks below the band gap of quantum wells, which are sensitive to the electric field [157]. However, due to screening, excitons usually disappear at the high carrier concentrations required for lasing.

5.1.5 Gain Suppression

Conventionally, the optical gain is considered independent of the photon density (linear gain). However, the optical gain tends to decrease at high photon densities S (gain suppression). This nonlinear gain effect is partially caused by the depletion of electrons at certain energy levels due to strong stimulated recombination (spectral hole burning). Gain suppression is commonly approximated as

$$g = \frac{g_0}{1 + \epsilon S},$$
(5.23)

where the nonlinear gain suppression coefficient ϵ is typically on the order of $10^{-17} \, \text{cm}^3$. In the case of multiple modes, the photon density S is the sum of the densities of all the modes.

5.2 Spontaneous Emission

The total spontaneous emission rate in bulk material is often approximated by Eq. (3.34) using the bimolecular recombination coefficient B. This simple equation represents the full spectrum of photons generated by spontaneous band-to-band recombination processes. The spontaneous emission spectrum itself can be given as

$$r_{\text{sp}}(h\nu) = \left(\frac{q^2 h}{2m_0^2 \varepsilon \varepsilon_0} \right) \left(\frac{1}{h\nu} \right) |M_{\text{b}}|^2 D_{\text{r}} D_{\text{opt}} f_{\text{c}} (1 - f_{\text{v}}),$$
(5.24)

and it is closely related to the gain spectrum in Eq. (5.1). In the case of quantum wells, all subbands need to be included. The transition strength $|M_{\text{b}}|^2$ is now averaged over all polarization directions. The emission rate is proportional to the

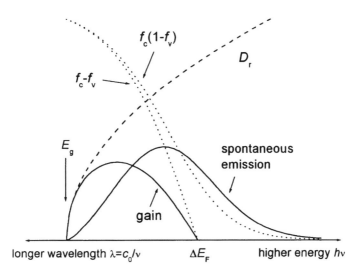

Figure 5.11: Schematic spectra of spontaneous emission and gain, respectively, without energy broadening for a bulk semiconductor (solid). The dotted lines give the Fermi factors for both cases. The dashed line illustrates the reduced density of states D_r.

reduced density of electron states $D_r(E)$ (see Eq. (5.2)) as well as to the density of photon states D_{opt}. For large optical cavities, the density of photon states becomes

$$D_{opt}(h\nu) = \frac{\varepsilon n_r}{\pi^2 \hbar^3 c^3}(h\nu)^2. \tag{5.25}$$

The Fermi factor $f_c(1 - f_v)$ gives the probability that the conduction band level is occupied and the valence band level is empty at the same time. It is different from the Fermi factor $f_c - f_v$ of the gain. Figure 5.11 illustrates the effect of the different Fermi factors on gain and spontaneous emission. The spontaneous emission spectrum peaks at higher photon energy, and it extends beyond $h\nu = \Delta E_F$. The final spontaneous emission spectrum $r_{spon}(h\nu)$ is obtained by including the transition energy broadening analogue to Eq. (5.19). This spectrum can be connected to the polarization averaged gain spectrum

$$\overline{g}(h\nu) = \frac{n_r}{c}n_{sp}^{-1}(h\nu)D_{opt}^{-1}(h\nu)r_{spon}(h\nu) \tag{5.26}$$

with the population inversion factor

$$n_{sp}(h\nu) = \frac{f_c(1 - f_v)}{f_c - f_v} = \left\{1 - \exp\left[\frac{h\nu - q\Delta E_F}{k_B T}\right]\right\}^{-1}. \tag{5.27}$$

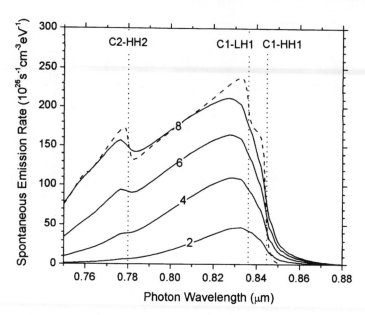

Figure 5.12: Spectra of the spontaneous emission rate for our 8-nm-thick GaAs/Al$_{0.15}$Ga$_{0.85}$As quantum well. The parameter gives the carrier concentration $n = p$ in 10^{18} cm^{-3}. Solid lines are for $\tau_s = 0.1$ ps; the dashed line is for $\tau_s = 0.5$ ps.

Figure 5.12 plots spontaneous emission spectra for our GaAs quantum well. As carrier inversion is not required here, spontaneous emission is present at all carrier concentrations. Transition energy broadening smoothens the spontaneous emission spectrum, and the scattering time τ_s can be extracted from measurements. The spontaneous emission spectrum is easier to measure than the gain spectrum, and it is often used for quantum well characterization. However, carrier concentrations are small and usually not exactly known in these measurements, generating some uncertainty in the extraction of broadening parameters.

Broadening does not affect the total spontaneous recombination rate, which is obtained by integration over all photon energies

$$R_{\text{spon}} = \int_0^\infty dE \, r_{\text{sp}}(E). \tag{5.28}$$

This integral is more accurate than Eq. (3.34), and it also applies to quantum wells.

Further Reading

- W. W. Chow and S. W. Koch, *Semiconductor Laser Fundamentals*, Springer-Verlag, Berlin, 1999.

- S. L. Chuang, *Physics of Optoelectronic Devices*, Wiley, New York, 1995.

- J. P. Loehr, *Physics of Strained Quantum Well Lasers*, Kluwer, Boston, 1998.

Chapter 6

Heat Generation and Dissipation

This chapter introduces the heat flux equation, as well as typical heat sources in semiconductor optoelectronic devices. For various semiconductors, the thermal conductivity of the crystal lattice and its heat capacity are listed. Interpolation formulas for the strong reduction of the thermal conductivity in semiconductor alloys are given.

6.1 Heat Flux Equation

Self-heating often limits the performance of optoelectronic devices. Heat is generated when carriers transfer part of their energy to the crystal lattice. In consequence, the thermal (vibrational) energy of the lattice rises, which is measured as an increase in its temperature, T_L. In this chapter, we assume a local thermal equilibrium between lattice and carriers with $T = T_L = T_n = T_p$ (cf. Section 3.9.1). Virtually all material properties like energy band gap and carrier mobility change with rising temperature. Within the crystal lattice, thermal energy is dissipated by traveling lattice vibrations. The smallest energy portions of lattice waves are called phonons, which can be treated like particles. Microscopic theories of lattice heat generation and dissipation are based on the phonon picture outlined in many solid-state textbooks, e.g., [46, 103, 158].

In practical device simulation, the main thermal parameters are thermal conductivity κ_L and specific heat C_L of the crystal lattice (Table 6.1). Electrons and holes also contribute to specific heat and thermal conductivity (cf. Section 3.9.1). However, those contributions are usually less than 1% of the total values [159]

$$C_{th} = C_L + C_n + C_p \tag{6.1}$$

$$\kappa_{th} = \kappa_L + \kappa_n + \kappa_p, \tag{6.2}$$

and they are neglected in the following. The lattice thermal conductivity κ_L controls the heat flux density (W/cm^2)

$$\vec{J}_{heat} = -\kappa_L \nabla T, \tag{6.3}$$

141

Table 6.1: Crystal Lattice Thermal Conductivity κ_L, Specific Heat C_L, Density ρ_L, Debye Temperature Θ_D, and Temperature Coefficient δ_κ at Room Temperature [1, 3, 6, 38, 46, 69]

Parameter Unit	κ_L (W/Kcm)	C_L (Ws/gK)	ρ_L (g/cm^3)	Θ_D (K)	δ_κ —
Si	1.31	0.703	2.328	648	−1.03
Ge	0.58	0.322	5.323	366	−1.4
GaAs	0.44	0.327	5.318	345	−1.25
InP	0.68	0.311	4.81	420	−1.5
AlAs	0.91	0.45	3.76	417	
GaSb	0.33	0.25	5.614	269	−1.1
AlSb	0.57		4.26	370	−1.2
InAs	0.27	0.352	5.667	248	−1.2
GaP	0.77	0.313	4.138	445	−1.3
AlP	1.30	0.477	2.4		
InSb	0.17	0.209	5.775	161	
GaN	1.30	0.49	6.15	600	−0.28
AlN	2.85	0.6	3.23	1150	−1.64
InN	0.45	0.32	6.81	660	
ZnS	0.27	0.47	4.075	315	−1.32
ZnSe	0.19	0.339	5.27	273	−1.12
ZnTe	0.18	0.264	5.636	225	−1.48
CdS	0.2	0.33	4.82	215	−1.14
CdSe	0.043	0.255	5.81	180	
CdTe	0.05	0.205	5.87	162	

Note. Nitride parameters are for wurtzite crystals, all others are for cubic lattices.

which follows the slope of the temperature distribution $T(\vec{r})$.[1] Conservation of energy requires that the temperature satisfy the heat flux equation

$$\rho_L C_L \frac{\partial T}{\partial t} = -\nabla \cdot \vec{J}_{\text{heat}} + H_{\text{heat}} \tag{6.4}$$

where ρ_L is the material's density and $H_{\text{heat}}(\vec{r}, t)$ is the heat power density (W/cm^3) generated by various sources. This equation relates the change in local temperature ($\partial T / \partial t$) to the local heat flux (in or out) and to the local heat generation.

---

[1]The energy balance model also includes convective heat transfer by carriers (cf. Eq. (3.67)).

All material parameters in Eq. (6.4) generally depend on the local position and on the temperature itself. Near room temperature [46]

$$C_L(T) = C_L(300\,\text{K})\frac{20 - (\Theta_D/T)^2}{20 - (\Theta_D/300\,\text{K})^2} \tag{6.5}$$

with Θ_D giving the Debye temperature, which itself is temperature dependent [3]. Temperature effects on the thermal conductivity near room temperature can be described by a power law

$$\kappa_L(T) = \kappa_L(300\,\text{K})\left(\frac{T}{300\,\text{K}}\right)^{\delta_\kappa}. \tag{6.6}$$

Over a wider temperature range, the relation $\kappa_L(T)$ is more complex as different scattering mechanisms dominate at different temperatures [46]. The material parameters of the above formulas are listed in Table 6.1.

The random distribution of alloy atoms in ternary or quaternary semiconductor compounds causes strong alloy scattering of phonons, which leads to a significant reduction in the thermal conductivity. The thermal conductivity of ternary alloys AB_xC_{1-x} can be estimated from binary values using

$$\frac{1}{\kappa_L(x)} = \frac{x}{\kappa_{AB}} + \frac{1-x}{\kappa_{AC}} + x(1-x)C_{ABC} \tag{6.7}$$

with the empirical bowing parameter C_{ABC} (Table 6.2). The same bowing parameters can be employed for quaternary alloys $A_xB_{1-x}C_yD_{1-y}$ [43, 160, 161]

$$\frac{1}{\kappa_L(x,y)} = \frac{xy}{\kappa_{AC}} + \frac{x(1-y)}{\kappa_{AD}} + \frac{(1-x)y}{\kappa_{BC}} + \frac{(1-x)(1-y)}{\kappa_{BD}}$$
$$+ x(1-x)\,[yC_{ABC} + (1-y)C_{ABD}]$$
$$+ y(1-y)\,[xC_{ACD} + (1-x)C_{BCD}]. \tag{6.8}$$

Alternatively, alloys of the type $AB_xC_yD_{1-x-y}$ are described by

$$\frac{1}{\kappa_L(x,y)} = \frac{x}{\kappa_{AB}} + \frac{y}{\kappa_{AC}} + \frac{1-x-y}{\kappa_{AD}} + xyC_{ABC}$$
$$+ x(1-x-y)C_{ABD} + y(1-x-y)C_{ACD}. \tag{6.9}$$

Results plotted in Fig. 6.1 illustrate that alloy scattering can reduce the thermal conductivity by more than one order of magnitude. The specific heat C_L of ternary or quaternary alloys is obtained by linear interpolation of binary values [69].

Table 6.2: Thermal Conductivity Bowing Parameter C_{ABC} (Km/W) in Eqs. (6.7), (6.8), and (6.9) for Ternary Alloys A(B,C) [43, 160]

Alloy	C_{ABC}	Alloy	C_{ABC}	Alloy	C_{ABC}
Al(As,Sb)	65	In(As,Sb)	80	(Al,Ga)As	30
Al(P,Sb)	101	In(P,Sb)	115	(Al,In)As	80
Al(P,As)	39	In(P,As)	34	(Ga,In)As	78.8
Ga(As,Sb)	63	(Al,Ga)P	30	(Al,Ga)Sb	34
Ga(P,Sb)	91	(Al,In)P	77	(Al,In)Sb	88
Ga(P,As)	21.6	(Ga,In)P	19.9	(Ga,In)Sb	68

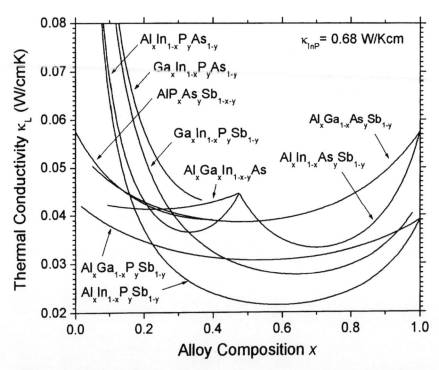

Figure 6.1: Room-temperature thermal conductivity of quaternary compounds lattice-matched to InP.

6.2 Heat Generation

Lattice heat is generated whenever physical processes transfer energy to the crystal lattice. According to differences in transfer mechanisms, heat sources can be separated into Joule heat, electron–hole recombination heat, Thomson heat, and heat from optical absorption. Here, we discuss these terms for the steady-state case, a more general discussion can be found in [159].

6.2.1 Joule Heat

The flow of carriers through a semiconductor is accompanied by frequent carrier scattering by phonons, leading to a continuing energy loss to the lattice. Carriers move from a higher electrostatic potential to a lower potential, and the corresponding energy difference is typically absorbed by the lattice as Joule heat

$$H_J = \frac{\vec{j}_n^2}{q\mu_n n} + \frac{\vec{j}_p^2}{q\mu_p p}, \tag{6.10}$$

which is proportional to the electrical resistance of the material (cf. Fig. 3.17c). The current density can be expressed by the slope of the quasi-Fermi levels, leading to the equivalent equation

$$H_J = -\frac{1}{q}(\vec{j}_n \nabla E_{Fn} + \vec{j}_p \nabla E_{Fp}). \tag{6.11}$$

6.2.2 Recombination Heat

When an electron–hole pair recombines, the energy is either transferred to a photon (light) or to a phonon (heat). The average heat released is proportional to the difference between the quasi-Fermi levels

$$H_R = R\,(E_{Fn} - E_{Fp}). \tag{6.12}$$

Besides defect recombination, the recombination rate $R = R_{SRH} + R_{Aug}$ also includes Auger recombination since the hot carriers generated during Auger recombination eventually lose their energy to phonons (cf. Fig. 3.10). Note that R is the net recombination rate, including thermal generation of carriers.

Spontaneous recombination may be included here since most of the photons emitted are absorbed by the semiconductor and eventually converted into heat. Stimulated emission of photons also leads to some heat generation as those photons are partially absorbed inside the device. However, heat generation from photon absorption is spread throughout the device and should better be treated in combination with optical calculations (Section 6.2.4).

Electron–hole recombination also causes a cooling of carriers above the Fermi level, which fill the now empty spots at the Fermi level. This contribution is related to the thermoelectric power P_p and P_n of holes and electrons, respectively, (cf. Section 3.9.1)

$$H_P = qRT(P_p - P_n). \qquad (6.13)$$

TqP_n and TqP_p are the average excess energies of carriers above the quasi-Fermi level. The excess energy is not unlikely to exceed the recombination energy $E_{Fn} - E_{Fp}$ [159].

6.2.3   Thomson Heat

The thermoelectric power (V/K) is a measure for the increase in average carrier excess energy with increasing temperature. It varies with the density of states, carrier concentration, and temperature (cf. Eqs. (3.72) and (3.73)). So-called Thomson heat is transferred between carriers and lattice as current flows along a gradient of the thermoelectric power

$$H_T = -qT(\vec{j}_n \nabla P_n + \vec{j}_p \nabla P_p). \qquad (6.14)$$

A dramatic example is the interface between different semiconductors. When entering a material with a lower conduction band edge, electrons suddenly exhibit excess kinetic energy (hot electrons) that is eventually dissipated to the lattice (cf. Fig 3.17d). In the opposite direction, electrons need to receive extra energy from the lattice to leave the quantum well. Therefore, Thomson heat can be positive or negative (cooling). Due to its close relation to the Peltier effect, Thomson heat is also referred to as Peltier heat (cf. Section 3.9.1).

6.2.4   Optical Absorption Heat

When optical waves penetrate a material, their energy can be partially or fully absorbed. The magnitude and the mechanism of absorption depends on the photon energy $h\nu$ (cf. Section 4.2.1). At low photon energies, the light is directly absorbed by the crystal lattice (*reststrahlen* region). At typical photon energies, absorption by free carriers dominates, which quickly dissipates the energy to the lattice due to very short intraband scattering times. Band-to-band absorption results in new electron–hole pairs (photon recycling), which may or may not generate heat by one of the above mechanisms. Corresponding to the overall device model, band-to-band absorption may be excluded as a direct heat source.

Using the appropriate optical absorption coefficient $\alpha_o(h\nu)$, the absorption-related heat power density can simply be given as

$$H_A = \alpha_o \Phi_{ph} \, h\nu \qquad (6.15)$$

with the photon flux density Φ_{ph} from Eq. (4.49). This formula is based on the assumption that electrons, if involved, transfer their absorption energy to the lattice without traveling.

6.3 Thermal Resistance

In practical device design, the internal temperature often needs to be known only in specific locations, e.g., within the active region of a laser diode. If the heat power P_{heat} (W) is generated in the same location, then the heat flux from that location to the heat sink can be characterized by a thermal resistance R_{th} (K/W), giving the temperature difference

$$\Delta T = R_{th} P_{heat} \tag{6.16}$$

between heat source and heat sink. Similar to the electrical resistance, the thermal resistance depends not only on material properties (thermal conductivity) but also on the device geometry.

The advantage of this approach is the thermal characterization of the device by one parameter R_{th} that usually can be measured. In cases with heat generation at different locations within the device, the single resistance R_{th} can be replaced by a thermal resistance network. In analogy to electrical circuits, simplified thermal models can be established this way. Thermal resistances are also employed to account for the heat flux outside the simulated device region (cf. Section 9.4).

6.4 Boundary Conditions

Similar to Section 3.5, three types of thermal boundary conditions are distinguished for the heat flux equation: *Dirichlet, Neumann*, and *mixed* conditions. The *Dirichlet* type simply specifies the temperature T_b at the boundary, and it is appropriate for device mounting onto a heat sink with small thermal resistance:

$$T_b = T_{sink} \tag{6.17}$$

with T_{sink} representing the ambient temperature. With larger external thermal resistance R_{th}^{sink}, the heat sink temperature T_{sink} becomes a function of the heat power. In this case, the *mixed* boundary condition for the surface-normal heat flux applies

$$\vec{v}\vec{J}_{heat} = \frac{T_b - T_{ext}}{A_{th} R_{th}^{sink}} \tag{6.18}$$

with the thermal contact area A_{th} and the external temperature T_{ext} on the other side of the heat sink. This boundary condition considers an unknown but uniform boundary temperature T_b. The boundary temperature becomes nonuniform with small simulated device regions or with similar thermal conductivities on both sides of the boundary. In that case, external regions must be included in the heat flux equation.

A similar boundary condition is used to describe convective heat removal from the device surface by moving air or liquids

$$\vec{v}\vec{J}_{\text{heat}} = h_c T_b - T_{\text{ext}} \qquad (6.19)$$

with the heat transfer coefficient h_c. For free convection of air, $h_c = 6 \ldots 30 \, \text{Wm}^{-2} \, \text{K}^{-1}$. Forced convection or liquids exhibit much higher coefficients. However, convective heat transfer is often negligible with optoelectronic devices since surface heating and surface area are relatively small. Radiative heat transfer is even less important.

Thus, most boundaries of a typical device exhibit negligible heat flux, which is described by the *Neumann* condition

$$\vec{v}\vec{J}_{\text{heat}} = 0. \qquad (6.20)$$

This is often the default thermal boundary condition in device simulation software. It applies to symmetry planes as well as to interfaces with air or vacuum.

Practical examples of self-heating are analyzed in Chapters 8 and 9 for vertical cavity lasers and GaN-based light emitters, respectively.

Further Reading

- B. Abeles, Lattice thermal conductivity of disordered semiconductor alloys at high temperatures, *Phys. Rev.*, **131**, 1906–1911 (1963).

- G. K. Wachutka, Rigorous thermodynamic treatment of heat generation and conduction in semiconductor device modeling, *IEEE Trans. Comput. Aided Design*, **9**, 1141–1149 (1990).

Part II
Devices

Chapter 7

Edge-Emitting Laser

This chapter gives an introduction to laser diodes, and it analyzes the physical mechanisms behind two key performance parameters: threshold current and slope efficiency. A common InP-based multiquantum well laser diode is used as example, and laser measurements are employed to calibrate material parameters in the simulation. The two-dimensional laser model self-consistently combines gain calculations with a drift–diffusion model and optical waveguide simulations. Degrading effects of ambient temperature elevation on the laser performance are investigated. Internal physical processes that are hardly accessible by experiments or by simple laser models are revealed.

7.1 Introduction

Traveling through a semiconductor, a single photon is able to generate an identical second photon by stimulating the recombination of an electron–hole pair. This photon multiplication is the key physical mechanism of lasing. The second photon exhibits the same wavelength and the same phase as the first photon, doubling the amplitude of their monochromatic wave. Subsequent repetition of this process leads to strong light amplification. However, the competing process is the absorption of photons by the generation of new electron–hole pairs. Stimulated emission prevails when more electrons are present at the conduction band level than at the valence band level (cf. Chap. 5). This carrier inversion is the first requirement of lasing and it is achieved at pn-junctions by providing conduction band electrons from the n-doped side and valence band holes from the p-doped side (Fig. 7.1). The photon energy is given by the band gap, which depends on the semiconductor material. Continuous current injection into the device leads to continuous stimulated emission of photons, but only if enough photons are constantly present in the device to trigger this process. Therefore, optical feedback and the confinement of photons in an optical resonator is the second basic requirement of lasing. In a Fabry–Perot laser, two reflecting facets at both ends of the optical waveguide are used for optical feedback (laser length L, facet power reflectivities R_1, R_2). The photon round trip gain is given by

$$G_{rt} = R_1 R_2 \exp[(\Gamma_o g - \alpha_i)L] = \exp[(\Gamma_o g - \alpha_i - \alpha_m)L]. \qquad (7.1)$$

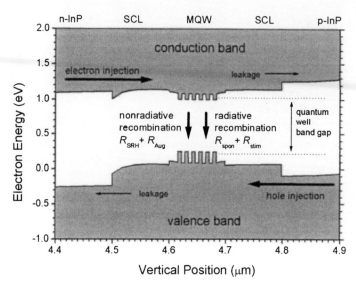

Figure 7.1: Band diagram of our InGaAsP/InP laser diode illustrating carrier transport processes (see text).

The quantum well gain g rises with increasing concentration of electrons and holes. The volume of the optical mode V_m is typically much larger than the active layer volume V_a; their ratio is given by the optical confinement factor

$$\Gamma_o = \frac{V_a}{V_m}. \tag{7.2}$$

For lasing, the modal gain $\Gamma_o g$ needs to compensate for internal optical loss α_i and mirror loss α_m. Lasing threshold is reached at $G_{rt} = 1$ with the threshold gain

$$\Gamma_o g_{th} = \alpha_i + \alpha_m. \tag{7.3}$$

Figure 7.1 illustrates the injection of electrons and holes into a multiquantum well (MQW) active region as well as different recombination processes. Crystal defects (SRH, Shockley–Read–Hall recombination) and the Auger process cause nonradiative recombination. Photons are generated by spontaneous and by stimulated recombination. Carriers may also leak out of the separate confinement layer (SCL). Vertical leakage is typically dominated by electrons and lateral leakage by ambipolar carrier diffusion within the quantum wells (Fig. 7.2). Carriers leaving the ridge waveguide region (gray area in Fig. 7.2) eventually recombine; however, they are considered leakage carriers in our analysis. Recombination mechanisms inside the waveguide region plus both the leakage currents add up to the total

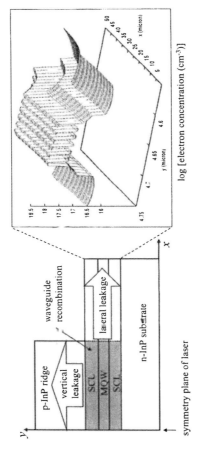

Figure 7.2: Left: Half cross section of the ridge–waveguide laser indicating leakage currents. Right: Electron distribution $n(x, y)$ in active layers.

153

injection current. All carriers that do not contribute to stimulated recombination are considered lost carriers. The threshold current I_{th} compensates for all the carrier losses at lasing threshold

$$I_{th} = I_r + I_v + I_l \qquad (7.4)$$

(I_r, recombination losses; I_v, vertical leakage; I_l, lateral leakage). Above threshold, the lasing power

$$P(I) = \eta_d \frac{\hbar\omega}{q}(I - I_{th}) = \eta_i \frac{\alpha_m}{\alpha_m + \alpha_i} \frac{\hbar\omega}{q}(I - I_{th}) \qquad (7.5)$$

rises almost linearly with increasing injection current I (Fig. 7.3, $\hbar\omega$, photon energy; q, elementary charge). Note that $P(I)$ stands for the total lasing power from both facets. In the case of a symmetrical Fabry–Perot laser, both facets emit the same power. The slope efficiency dP/dI is proportional to the differential quantum efficiency η_d. It depends on the internal optical loss α_i, on the mirror loss

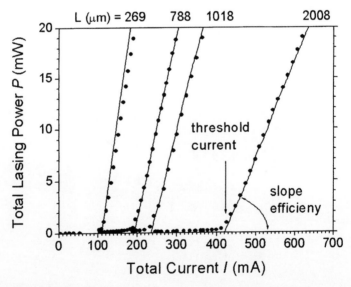

Figure 7.3: Light–current (LI) characteristics in pulsed operation at room temperature with different cavity lengths L (dots, measurement; lines, simulation).

α_m, and on the injection efficiency

$$\eta_i = \frac{\Delta I_{stim}}{\Delta I} = \frac{\Delta I_{stim}}{\Delta I_{stim} + \Delta I_r + \Delta I_v + \Delta I_l}, \tag{7.6}$$

which gives the fraction of the total current increment ΔI that results in stimulated emission of photons. The parameter η_i is less than unity if parts of ΔI are consumed by other recombination processes (ΔI_r) or by leakage (ΔI_v, ΔI_l). The corresponding differential efficiencies are given as

$$\eta_i = \eta_l \times \eta_v \times \eta_r \tag{7.7}$$

$$= \frac{\Delta I_{stim} + \Delta I_r + \Delta I_v}{\Delta I_{stim} + \Delta I_r + \Delta I_v + \Delta I_l} \times \frac{\Delta I_{stim} + \Delta I_r}{\Delta I_{stim} + \Delta I_r + \Delta I_v} \times \frac{\Delta I_{stim}}{\Delta I_{stim} + \Delta I_r}. \tag{7.8}$$

These efficiencies can also be understood as probabilities. An additional electron injected above the threshold has the probability η_l of remaining within the ridge region. It has the probability $\eta_l \eta_v$ of recombining within the waveguide region. The efficiency η_r gives the ratio of the stimulated recombination increment to the total recombination increment within the waveguide region (including the MQW). These contributions to the slope efficiency will be further analyzed in Sections 7.4 and 7.5.

The simulation example in this chapter is a typical InGaAsP multiquantum well laser diode grown on InP [162]. The electron energy band diagram is given in Fig. 7.1, and the layer material are listed in Table 7.1. The MQW active region

Table 7.1: Layer Materials and Room Temperature Parameters of the MQW Fabry–Perot Laser

Parameter Unit	d (μm)	N_{dop} ($1/cm^3$)	μ (cm^2/Vs)	n_r —
p-InP (cladding)	1.86	4×10^{17}	100	3.17
p-InP (doping offset)	0.14	—	150	3.17
$In_{0.83}Ga_{0.17}As_{0.37}P_{0.63}$ (SCL)	0.1	—	100	3.28
$In_{0.76}Ga_{0.24}As_{0.79}P_{0.21}$ (QW)	0.0064	—	100	3.42
$In_{0.71}Ga_{0.29}As_{0.55}P_{0.45}$ (barrier)	0.0055	—	100	3.34
$In_{0.83}Ga_{0.17}As_{0.37}P_{0.63}$ (SCL)	0.1	—	100	3.28
n-InP (cladding)	1.5	8×10^{17}	4500	3.17
n-InP (substrate)	300	6×10^{18}	4500	3.17

Note. d, layer thickness; N_{dop}, doping density; μ, majority carrier mobility; n_r, refractive index at 1.55 μm wavelength.

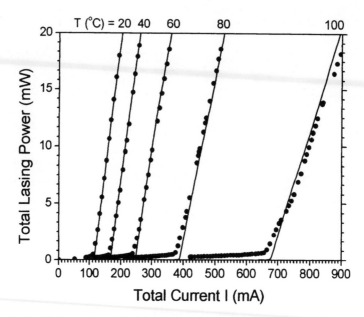

Figure 7.4: Light–current (LI) characteristics in pulsed operation with the ambient temperature given as a parameter (269 μm cavity length; dots, measurement; lines, simulation).

consists of six 6.4-nm-thick 1% compressively strained InGaAsP quantum wells with 1.55 μm emission wavelength. The 5.5-nm-thick barriers exhibit slight tensile strain (0.04%). The first and last barriers are 17 nm wide. The MQW stack is sandwiched between undoped InGaAsP separate confinement layers (SCLs), which act as a waveguide. On the p-side of the structure, the first 140 nm of the InP cladding layer are not intentionally doped and the remainder is Zn doped. The width of the p-InP ridge is 57 μm. Facet cleaving gives Fabry–Perot resonators with different cavity lengths L. Measured light–current characteristics are shown in Figs. 7.3 and 7.4.

7.2 Models and Material Parameters

Laser diodes represent a complex interaction of electronic, thermal, and optical processes. We here employ the PICS3D software package, which self-consistently combines carrier transport, heat flux, optical gain computation, and wave guiding. However, the model includes many material parameters that are not exactly known. Critical parameters need to be calibrated using measured laser characteristics as explained in the following paragraphs. Such a calibration procedure is of paramount importance for drawing realistic conclusions from the simulation. Simultaneous reproduction of several measurements is often necessary for analyzing the relative importance of different physical mechanisms. For example, agreement with the measured threshold current is obtained by fitting the Auger recombination parameter or by fitting the absorption coefficient. The correct balance between both mechanisms can be found by simultaneous reproduction of the measured slope efficiency. However, the number of uncertain material parameters should be kept as small as possible by simplifying the experimental situation simulated. Laser operation with very short current pulses, for instance, avoids self-heating of the device, and it makes it possible to exclude heat flux from the model. We here use pulsed LI characteristics measured at different laser lengths (Fig. 7.3) [163] and at different ambient temperatures (Fig. 7.4) [164].

7.2.1 Drift–Diffusion Model

The drift–diffusion model of carrier transport includes Fermi statistics and thermionic emission at heterobarriers. This process is mainly controlled by the offset of the conduction band (ΔE_c) and the valence band (ΔE_v) at the heterobarrier ($E_g = 1.35 - 0.775y + 0.149y^2$ (eV) for $In_{1-x}Ga_xAs_yP_{1-y}$ lattice matched to InP with $x = 0.1886/(0.4192 - 0.0141y)$). We find best agreement with high-temperature LI measurements by using a band offset ratio of $\Delta E_c/\Delta E_v = 0.4/0.6$, which is typical for the InGaAsP/InP system (cf. Section 2.5). In this steady-state analysis, carrier transitions between confined and unconfined quantum well states are not treated explicitly. Bulk carriers are assumed to fill up the quantum states without delay, leading to single quasi-Fermi levels. Dynamic laser modeling considers characteristic delay times for such transitions [165].

Lateral leakage is small in our broad-area laser. LI and current–voltage measurements at different ridge widths can be used to identify lateral leakage [166] and to calibrate critical parameters like the hole mobility [167]. Within passive layers, a spontaneous emission parameter of $B = 10^{-10}$ cm^{3}s^{-1} is assumed. The spontaneous recombination rate in quantum wells is much larger than in passive layers, and it is calculated self-consistently from energy band structure and Fermi distribution including temperature effects (cf. Section 5.2). Typical numbers are assumed for the Shockley–Read–Hall (SRH) recombination lifetime of electrons and holes with 20 ns inside the active region and 100 ns elsewhere. Due to the small quantum well band gap, Auger recombination causes the strongest carrier losses in long-wavelength lasers. Various theoretical and experimental investigations of this mechanism have been published [168, 169, 170]. By definition, the experimental Auger parameter C is somewhat different from the theoretical parameters C_n and C_p used in calculations of the local Auger recombination rate $(C_n n + C_p p)(np - n_i^2)$ (cf. Section 3.7). For our type of laser with dominating valence band Auger recombination ($C_n = 0$), measured temperature effects can be reproduced using

$$C_p = C_o \exp\left[-\frac{E_a}{k_B T}\right] \qquad (7.9)$$

with an activation energy of $E_a = 60\,\mathrm{meV}$ [170]. Our fit to experimental LI characteristics leads to an Auger parameter of $C_p = 1.6 \times 10^{-28}\,\mathrm{cm^6 s^{-1}}$ at room temperature (Fig. 7.4).

7.2.2   Gain Model

For our strained quantum wells, the conduction bands are assumed to be parabolic and the nonparabolic valence bands are computed by the two-band $\vec{k} \cdot \vec{p}$ method including valence band mixing (cf. Section 2.2.1). The local optical gain is calculated self-consistently from the local Fermi distribution of carriers at each bias point of the LI curve. A Lorentzian broadening function is used with 0.1 ps intraband scattering time. Band gap renormalization due to carrier–carrier interaction is considered as

$$\Delta E_{\mathrm{g}} = -\xi \left(\frac{n+p}{2}\right)^{1/3} \qquad (7.10)$$

with $\xi = 10^{-8}$ eV cm. The thermal band gap shrinkage parameter $dE_{\mathrm{g}}/dT = -0.28\,\mathrm{meV/K}$ is extracted from the measured thermal shift of the lasing wavelength. Temperature effects on the calculated quantum well gain are shown in Figs. 7.5 and 7.6, including intervalence band absorption (see below). The red-shift of the gain peak wavelength with higher temperature is attributed to the shrinking band gap. The peak gain decreases substantially due to the wider spreading of the Fermi distribution of carriers. To maintain the required threshold gain with rising temperature, carrier concentration and injection current need to be increased. Section 7.5 shows that this mechanism triggers the strong temperature sensitivity of the threshold current.

7.2.3   Optical Model

The refractive index profile of our structure is listed in Table 7.1 and plotted in Fig. 7.7 together with the optical intensity. The optical confinement factor is $\Gamma_o = 0.074$. The local absorption coefficient is proportional to the concentration of electrons and holes

$$\alpha_o = \alpha_b + k_n n + k_p p . \qquad (7.11)$$

The constant background loss coefficient α_b represents carrier-concentration independent mechanisms like photon scattering at defects. Free-carrier absorption

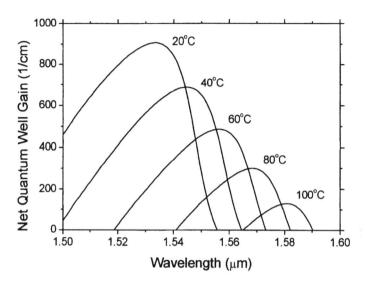

Figure 7.5: Net quantum well gain spectrum at different temperatures with constant carrier concentration ($n = p = 2 \times 10^{18}$ cm^{-3}).

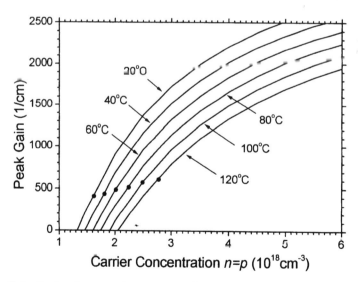

Figure 7.6: Peak gain vs carrier concentration with the ambient temperature as a parameter (dots, lasing threshold).

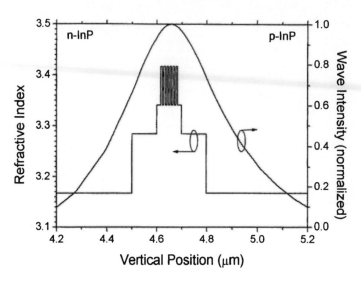

Figure 7.7: Vertical profile of refractive index and optical intensity.

due to electrons is known to be very small in 1.55-µm InGaAsP/InP lasers ($k_n \approx 10^{-18}$ cm^2) [74]. Intervalenceband absorption (IVBA) is usually considered the dominant absorption mechanism in 1.55-µm lasers. It is roughly proportional to the hole concentration. Absorption measurements at 1.55 µm wavelength give $k_p = 20\dots60 \times 10^{-18}$ cm^2 for bulk material (InGaAsP exhibits higher values than InP) [171, 172, 173]. It is difficult to accurately measure this parameter within quantum wells and only few experimental investigations of k_p can be found in the literature on 1.55-µm MQW structures [173, 174]. IVBA is often believed to weaken with compressive strain but some investigations suggest otherwise [175, 176]. Our fit to LI measurements at different cavity lengths gives $k_p = 82 \times 10^{-18}$ cm^2 (Fig. 7.3). This number is smaller than that measured with unstrained quantum wells (140×10^{-18} cm^2) and larger than that with more strongly strained quantum wells (35×10^{-18} cm^2 with 1.2% strain) [173]. However, using k_p as a fitted parameter helps to compensate for any inaccuracy of the gain calculation. Both IVBA and band-to-band transitions contribute to the net quantum well gain given in Figs. 7.5 and 7.6.

The internal modal absorption coefficient α_i is obtained by spatial integration of the local absorption α_o, weighted by the optical intensity. Alternatively, α_i could also be used as a fitted parameter without detailed absorption assessment for each layer. However, this simple approach would not reflect temperature effects on $\alpha_i(T)$, which are related to the local carrier concentration (see Section 7.5). For our lasers, vanishing background loss ($\alpha_b = 0$) and R = 0.28 facet power reflectance give best agreement with LI measurements at different laser lengths in Fig. 7.3. The mirror loss coefficient is $\alpha_m = 47\,\text{cm}^{-1}$ for a cavity length of $L = 269\,\mu\text{m}$.

7.3 Cavity Length Effects on Loss Parameters

The loss parameters α_i and η_i are typically extracted from LI characteristics measured at different cavity length L. Plotting the inverse slope efficiency as

$$\frac{1}{\eta_d} = \frac{1}{\eta_i}\left\{\frac{-2\alpha_i}{\ln(R_1 R_2)}L + 1\right\} \tag{7.12}$$

delivers α_i and η_i by linear regression of measured data. This widely used linear method is based on the assumption that η_i and α_i do not change with the laser length. This assumption is questionable for the following reasons.

Especially in long-wavelength lasers, absorption is mainly caused by free carriers (intraband transitions) and by intervalence band absorption. In both cases, the absorption coefficient rises proportionally to the carrier concentration. The injection efficiency η_i is influenced by the lateral spreading of carriers, carrier escape from the active region, and recombination losses within the active layers, which all depend on the carrier concentration. The mirror loss coefficient α_m rises with a shorter cavity length, requiring higher gain and more carriers in the quantum well. The higher carrier concentration causes the internal absorption to increase and the differential internal efficiency to decrease.

The functions $\alpha_i(L)$ and $\eta_i(L)$ are hard to extract experimentally but they can be obtained from our simulation (Fig. 7.0). Based on the good agreement with the measurements (Fig. 7.3), we first use Eq. (7.12) to extract both parameters from LI characteristics simulated with different laser lengths. This results in $\alpha_i = 14\,\text{cm}^{-1}$ and $\eta_i = 0.67$. Secondly, we extract the actual functions $\alpha_i(L)$ and $\eta_i(L)$ as follows. The internal absorption coefficient is obtained by spatial integration of the local absorption weighted by the optical intensity. As expected, the internal loss increases with shorter lasers. For comparison, the result of the linear method is shown as dotted line in Fig. 7.8, and it is only slightly smaller than the actual internal loss in long lasers. However, the linear method underestimates the actual absorption in short lasers substantially. The error is about 30% for our lasers with $L = 269\,\mu\text{m}$. The injection efficiency is extracted by spatial integration of the stimulated recombination rate at two different bias points above the threshold. For long lasers, the resulting function $\eta_i(L)$ is slightly above the linear result in Fig. 7.8. The efficiency $\eta_i(L)$ drops significantly with a shorter cavity length; however, it is close to the linear result for our lasers with $L = 269\,\mu\text{m}$. In summary, the linear method used in experiments is not accurate for short lasers.

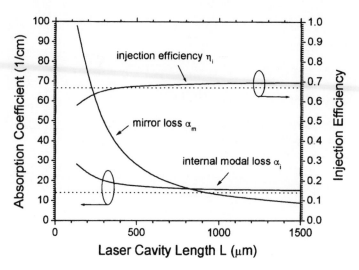

Figure 7.8: Mirror loss α_m, internal optical loss α_i, and injection efficiency η_i as a function of laser length (solid lines). Dotted lines give the result of the linear method used in experiments.

7.4 Slope Efficiency Limitations

The slope efficiency is one of the most important performance parameters of laser diodes. It gives the percentage of electrons injected above the threshold that contributes photons to the emitted laser beam. This efficiency is restricted by carrier losses (η_i) and by photon losses (α_i). The injection efficiency η_i is equal to the fraction of current above the threshold that results in stimulated emission of photons. Low injection efficiency η_i is usually attributed to vertical electron leakage, which is known to escalate at higher temperatures. Additional electron stopper layers introducing a barrier in the conduction band have been used successfully to enhance the high-temperature performance of long-wavelength lasers [177]. For our MQW laser, $\eta_i = 67\%$ is measured at room temperature, which suggests strong electron leakage. Thus, a second laser structure with an additional InGaP electron stopper layer between SCL and p-InP was fabricated [162]. The stopper layer has a conduction band offset to InP of $\Delta E_c \approx 50\,\text{meV}$ but it does not hinder hole transport. However, no significant change in the laser characteristics is measured at room temperature, indicating negligible vertical leakage [162]. Since both η_v and η_l are close to 100% in these broad-area lasers, the attention is drawn to recombination losses (η_r) to account for the reduced internal efficiency measured (cf. Eq. (7.7)). Commonly, the quantum well carrier concentration is considered constant above the threshold leading to $\eta_r = 100\%$ (Fermi level pinning) [39].

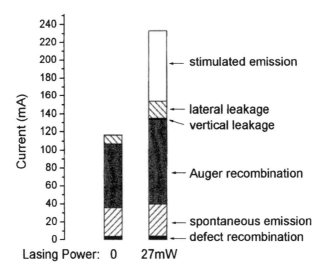

Figure 7.9: Contribution of carrier losses to the total current at threshold and at 27 mW output power ($T = 20°C$, $L = 269\,\mu m$).

This phenomenon is analyzed by evaluating carrier loss mechanisms in our laser simulation at room temperature (Fig. 7.9). Electron leakage from the active region can be identified as minority carrier current in the p-InP cladding layer. At threshold, far less than 1% of the total electron current leaks into the p-cladding; i.e., adding an InGaP barrier cannot have any significant effect at room temperature. Hole leakage is even smaller. Lateral leakage accounts for about 8% of the threshold current. Carrier losses due to Auger recombination, spontaneous emission, and defect recombination are analyzed by integrating the recombination rates at different bias points. At threshold, 61% of the total current feeds Auger recombination, 27% spontaneous emission, and 3% defect recombination. Figure 7.9 shows that all those carrier losses grow above the threshold. This growth contradicts the common belief that the quantum well carrier concentration does not change above the threshold [39]. However, even with constant total carrier number, the carrier distribution among the quantum wells changes with increasing bias [163]. In our MQW, electrons can move more easily across the MQW than holes so that p-side quantum wells are occupied by more carriers than n-side quantum wells. This trend is enhanced with rising bias. Figure 7.10 illustrates the effects on Auger recombination. The increment of Auger recombination within the more populated p-side quantum wells is larger than the decrement in the less populated n-side quantum wells. Auger recombination and spontaneous emission exhibit a superlinear dependence on the local carrier concentration, and increasing carrier nonuniformity causes stronger total recombination. SRH recombination

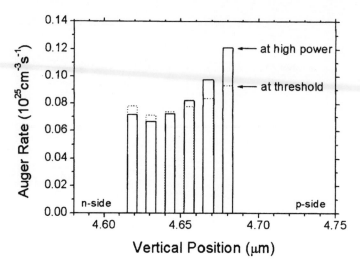

Figure 7.10: MQW Auger recombination at two different power levels ($T = 20°C$, $L = 269\,\mu m$).

grows linearly with the local carrier concentration, and it reflects the change in average carrier concentration. The net increase of recombination losses results in $\eta_r = 74\%$ for our device. Using Eq. (7.8), we extract the leakage contributions $\eta_v = 99\%$ and $\eta_l = 93\%$. The total injection efficiency $\eta_i = 68\%$ is very close to the measured number, and it is dominated by recombination losses.

7.5 Temperature Effects on Laser Performance

The performance of long-wavelength InGaAsP/InP laser diodes is known to be strongly temperature dependent. Self-heating or ambient temperature elevation causes the threshold current to increase and the slope efficiency to decrease. The physical mechanisms dominating the temperature sensitivity are still under discussion. In recent years, this discussion includes Auger recombination [178], IVBA [167], thermionic carrier emission out of the active region [179], lateral carrier spreading [180], passive layer absorption [181], spontaneous recombination within passive layers [182], and optical gain reductions [168, 183]. All of these mechanisms should be considered simultaneously in the analysis of measurements. One-sided models lead to one-sided interpretations of experiments and contribute to the controversy in this field. We here demonstrate the analysis of temperature effects by considering all of the above physical mechanisms and their self-consistent interaction.

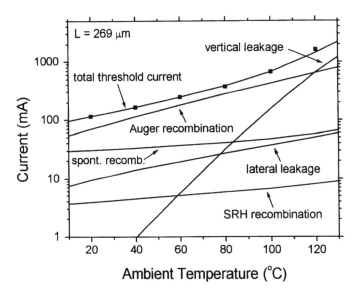

Figure 7.11: Pulsed threshold current and its components as a function of the ambient temperature (dots, measurement; lines, simulation).

Figure 7.11 plots the calculated threshold current and its components as a function of temperature. All recombination currents are obtained by spatial integration over the waveguide region (cf. Fig. 7.2). Carriers leaving the waveguide region in a lateral or vertical direction constitute leakage currents. At lower temperatures, the strongest contribution to the threshold current comes from Auger recombination, followed by spontaneous emission, lateral leakage current, and SRH recombination. Vertical carrier leakage is negligible at room temperature but it increases strongly and becomes the dominant carrier loss mechanism above 120°C. Thus, the temperature sensitivity of the threshold current is dominated by Auger recombination at lower temperatures and by vertical leakage at higher temperatures.

The escalation of carrier losses with higher temperature is related to the increasing carrier concentration (Fig. 7.12), which is mainly triggered by the gain reduction with higher temperature. This becomes clear from Fig. 7.13. With constant gain, the temperature sensitivity of the threshold current would be very small, despite Auger recombination. Thus, the reduced gain is the trigger mechanism for increasing carrier losses. Figure 7.13 also shows the effects of other model variations. Excluding Auger recombination reduces both the threshold current and its temperature sensitivity substantially. Including an electron stopper layer affects the threshold current only at high temperatures, when vertical leakage becomes relevant [162].

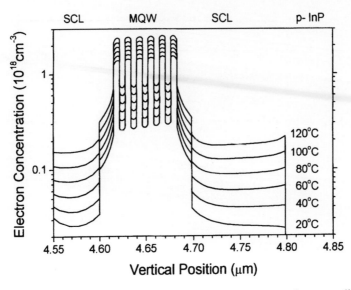

Figure 7.12: Vertical profile of the electron concentration at different temperatures.

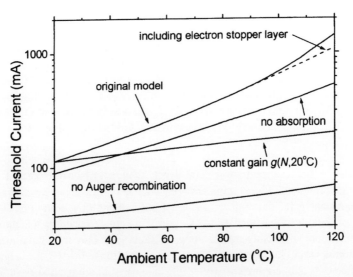

Figure 7.13: Threshold current as a function of ambient temperature with model variations.

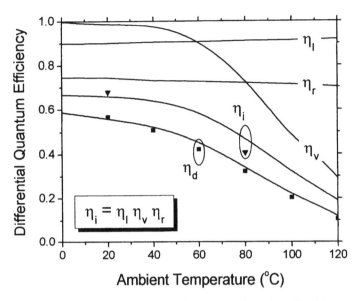

Figure 7.14: Components of the slope efficiency as a function of ambient temperature (η_d, differential quantum efficiency; η_i, injection efficiency; η_l, lateral leakage contribution; η_v, vertical leakage contribution; η_r, recombination contribution; dots, measurements; lines, simulation).

Components of the slope efficiency are plotted in Fig. 7.14 as a function of temperature (cf. Section 7.1). At room temperature, the injection efficiency η_i is dominated by recombination (η_r). Electron leakage is of minor importance. Above 80°C, vertical electron leakage causes the highest differential carrier loss (η_v). The recombination loss increment ΔI_r rises little with higher temperature. This is related to the more uniform hole distribution, which compensates for enhanced Auger recombination. Due to a lower carrier mobility, lateral leakage (η_l) decreases slightly, thereby reducing the temperature sensitivity of the slope efficiency, as observed in [166].

Finally, we analyze temperature effects on the internal optical loss $\alpha_i(T)$. Absorption is governed by the concentration of holes, which is highest in quantum wells. At room temperature, 64% of the modal absorption occurs within quantum wells and 10% within barriers and SCLs (Fig. 7.15). The remaining 26% originate in the p-InP cladding layer, which is occupied by a considerable part of the guided wave (cf. Fig. 7.7). At 120°C, quantum wells cause 60% of the modal absorption, whereas the contribution of barriers and SCLs rises to 24%. The total absorption doubles within this temperature range. Absorption by unconfined carriers rises strongly with temperature elevation; however, passive layer absorption

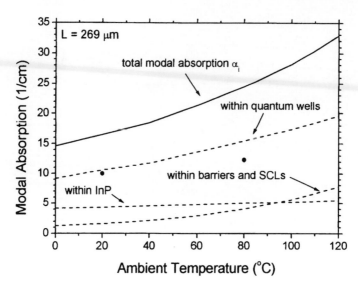

Figure 7.15: Internal absorption coefficient α_i and its components as a function of temperature (dots, α_i obtained by the linear method).

does not dominate internal optical losses as suggested by more simple models [170]. The dots in Fig. 7.15 give experimental results for α_i, which are smaller than calculated. As discussed in Section 7.3, this disagreement is an inherent problem of the experimental method with short lasers and it does not invalidate our simulation.

Photon losses have less effect on the slope efficiency η_d than carrier losses. From 20 to 120°C, the optical efficiency $\alpha_m/(\alpha_m+\alpha_i)$ only decreases from 74% to 56% in our case. The temperature sensitivity of the threshold current is also hardly affected by absorption (Fig. 7.13). However, the threshold current at 120°C is almost three times smaller without absorption, mainly due to reduced vertical leakage.

This chapter analyzed pulsed laser operation without self-heating. Continuous-wave operation of different types of laser diodes, including heat generation and heat dissipation, is investigated in the following two chapters.

Further Reading

- L. A. Coldren and S. W. Corzine, *Diode Lasers and Photonic Integrated Circuits*, Wiley, New York, 1995.

- G. P. Agrawal and N. K. Dutta, *Semiconductor Lasers*, van Nostrand Reinhold, New York, 1993.

- P. Zory (Ed.), *Quantum Well Lasers*, Academic Press, San Diego, 1993.

- J. Carroll, J. Whiteaway, and D. Plumb, *Distributed Feedback Semiconductor Lasers*, SPIE Press, Bellingham, WA, 1998.

- E. Kapon (Ed.), *Semiconductor Lasers*, Academic Press, San Diego, 1999.

Chapter 8

Vertical-Cavity Laser

Vertical-cavity surface-emitting lasers (VCSELs) are briefly introduced. Using wafer-bonded long-wavelength VCSELs as an example, we demonstrate how advanced simulations are utilized in combination with measurements to design and analyze real devices. Section 8.2 describes the device structure and Section 8.3 outlines the numerical VCSEL model. The following sections give details of the electrical, thermal, optical, and gain analysis. Measurements are employed to provide unknown device parameters to the simulation.

8.1 Introduction

In VCSELs, the optical cavity is formed by mirrors above and below the active region (Fig. 8.1). The laser light propagates in the vertical direction and typically exhibits a circular beam shape, ideal for coupling into optical fibers. Internally, the light passes the active layers in the vertical direction; i.e., gain is provided over a short distance only and the amplification per photon round trip is small. Therefore, the mirrors need to be highly reflective so that photons make many round trips before they are emitted. The high reflectivity is provided by distributed Bragg reflectors (DBRs) with two alternating layers having high refractive index contrast. With quarter-wavelength layer thickness, the reflected waves from all DBR inter faces add up constructively, allowing for DBR reflectances above 99%. Dielectric DBRs like Si/SiO_2 provide the highest refractive index contrast, and a few layer pairs are sufficient for high mirror reflectivity. However, semiconductor DBRs are often prefered in order to inject the current vertically through the mirror into the active region. AlAs/GaAs DBRs give a relatively high reflectivity, but the large band edge offset at the interfaces causes a high electrical resistance, especially in p-doped DBRs. Therefore, AlGaAs layers as well as sophisticated compositional grading and doping schemes are frequently employed at the interfaces. The electrical resistance and other heat sources have the potential to cause strong self-heating of VCSELs so that good thermal conductivity is another essential requirement of VCSEL design. For detailed reviews on VCSELs, see [184, 185].

8.2 Long-Wavelength Vertical-Cavity Lasers

Long-wavelength VCSELs (1.3–$1.6\,\mu$m) are currently of high interest for applications in fiberoptic communication systems. Compared to their in-plane

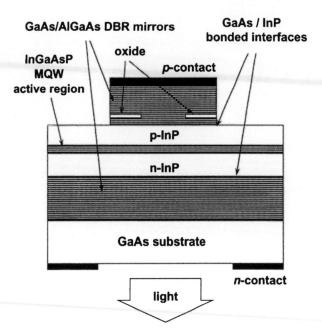

Figure 8.1: Schematic structure of a long-wavelength wafer-bonded vertical-cavity laser.

counterparts, VCSELs offer several advantages, including high fiber-coupling efficiency, low power consumption, and low-cost wafer-scale fabrication and testing. In contrast to the rapid development of GaAs-based VCSELs emitting at shorter wavelengths, the performance of InP-based long-wavelength devices is severely limited by disadvantageous material properties of the traditional InGaAsP semiconductor system at long wavelengths. With the lower band gap of the InGaAsP active region, Auger recombination causes higher nonradiative carrier losses (cf. Fig. 3.12). With lower photon energy, free-carrier and intervalence band absorption (IVBA) lead to enhanced optical losses. InGaAsP/InP DBRs only allow for a small variation of the refractive index, which is about half the variation possible in AlGaAs. To obtain high DBR reflectances, a large number of mirror layers must be grown causing significant diffraction loss [186]. InGaAsP also exhibits a low thermal conductivity due to the alloy scattering of phonons (cf. Section 6.1). An InGaAsP/InP DBR blocks the thermal flux to the stage and it leads to a strong increase of the active region temperature in continuous-wave (CW) operation.

For overcoming those limitations, several advanced concepts of long-wavelength VCSELs have been developed [187, 188]. One of the most successful concepts thus far is the utilization of InP/GaAs wafer bonding [189]. In this way, traditional InP-based active regions can be combined with superior GaAs-based mirrors and record-high lasing temperatures have been achieved [190, 191]. A wafer-bonded 1.55-μm VCSEL is shown schematically in Fig. 8.1 [192]. In this bottom-emitting device, 30 periods of GaAs/Al$_{0.67}$Ga$_{0.33}$As form the top DBR pillar, which is covered by a metal contact on a GaAs phase-matching layer to enhance reflectivity. A ring contact can be used for top emission. The mirror absorption is kept small by using relatively low p-doping (Table 8.1). The layer interfaces are parabolically graded to reduce the interface electrical resistance [193]. Close to the bonded interface, the mirror contains a 20-nm-thin Al$_{0.98}$Ga$_{0.33}$As layer for lateral oxidation, which is used for electrical and optical confinement. The InGaAsP multi quantum well (MQW) active region consists of 7 quantum wells with about 1% compressive strain and strain-compensating barriers. It is sandwiched between InP spacer layers, which are about 300 nm thick to give a total cavity thickness of three half-wavelengths. The bottom 28-period GaAs/AlAs DBR is pulse doped at all interfaces, in addition to uniform silicon doping. Binary layers are used for high reflectivity and good thermal conductivity. Several variations of this structure, with and without oxide confinement, are investigated in the following.

Table 8.1: Layer Materials and Parameters of the Double-Bonded VCSEL

Parameter Unit	d (μm)	N_{dop} (1/cm^3)	μ_z/μ_r (cm^2/Vs)	n_r —	α_0 (1/cm)	$\kappa_{\text{L}z}/\kappa_{\text{L}r}$ (W/cmK)
p-Al$_{0.67}$Ga$_{0.33}$As (DBR)	0.127	4×10^{17}	0.12^a/220	3.05	25^b	0.10/0.12
p-GaAs (DBR)	0.115	4×10^{17}	0.12^a/220	3.38	25^b	0.10/0.12
p-InP (spacer)	0.210	1×10^{18}	30	3.17	24	0.68
p-InP (spacer)	0.100	1×10^{16}	150	3.17	0.24	0.68
In$_{0.76}$Ga$_{0.24}$As$_{0.82}$P$_{0.18}$ (QW)	0.0055	—	100	3.6	54	0.043
In$_{0.48}$Ga$_{0.52}$As$_{0.82}$P$_{0.18}$ (barrier)	0.008	—	100	3.4	54	0.043
n-InP (spacer)	0.310	5×10^{18}	4600	3.15	8	0.68
n-GaAs (DBR)	0.115	1×10^{18}	310/6200	3.38	6	0.20/0.23
n-AlAs (DBR)	0.134	1×10^{18}	310/6200	2.89	3	0.20/0.23
n-GaAs (substrate)	450	5×10^{18}	8000	3.38	5.8	0.44

Note. d, layer thickness; N_{dop}, doping density; μ_z/μ_r, majority carrier mobility (vertical/lateral), n_r, refractive index; α_0, absorption coefficient; $\kappa_{\text{L}z}/\kappa_{\text{L}r}$ - lattice thermal conductivity (vertical/lateral).
^{a}For 1 kA/cm^2 current density (see Eq. (8.1)),
^{b}For 12 μm pillar diameter (see Eq. (8.2)).

8.3   Model and Parameters

The complexity and the interaction of electrical, thermal, and optical processes in VCSELs require advanced numerical models for device analysis and optimization. In order to analyze VCSEL performance near threshold, we use a comprehensive two-dimensional model that has produced excellent agreement with a variety of measurements [167, 194, 195]. Carrier transport is simulated using the drift–diffusion model. The heat flux equation is included to address self-heating effects. Gain calculations are based on 4×4 $\vec{k} \cdot \vec{p}$ band structure computations for the strained quantum wells. The transmission matrix method is employed for optical simulation including temperature effects on layer thickness, absorption, and refractive index. The lateral optical modes are given by Bessel functions.

Such a comprehensive model includes various material parameters, some of which are not exactly known. An important first step of every device simulation is to identify critical parameters and to calibrate them using measured device characteristics. Table 8.1 lists some of those parameters, the calibration procedures are discussed in the following sections. The measurements include variations of threshold current and slope efficiency with DBR diameter and with temperature (Figs. 8.2, 8.3) [167]. These pulsed measurements were taken on the first long-wavelength VCSELs that showed continuous-wave lasing above room

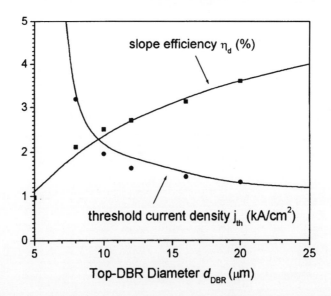

Figure 8.2: Threshold current density and slope efficiency as a function of top DBR diameter: dots, measurement; lines, simulation (pulsed operation at room temperature, no oxide layer).

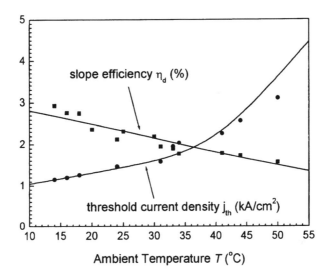

Figure 8.3: Threshold current density and slope efficiency as a function of temperature; dots: measurement, lines: simulation (pulsed operation, no oxide layer, top DBR diameter = $12 \, \mu m$).

temperature [196]. With a lower aperture diameter, lateral leakage currents and optical confinement losses increase, leading to a rising threshold current density and falling slope efficiency. Similar tendencies are observed with higher temperature, since stronger pumping is required for lasing, resulting in higher carrier losses (cf. Chap. 7). Calibration of model parameters related to these mechanisms gives good agreement between measurements and simulations.

8.4   Carrier Transport Effects

The carrier transport through the double-bonded VCSEL is affected by a large number of heterointerfaces. One of the most critical interfaces is the p-side bonded junction. Ideally, this interface between GaAs and InP exhibits only a small band offset [197]. However, in some bonded VCSELs, positively charged defects were found to create a large potential barrier in the valence band [198]. Figure 8.4 shows the band diagram along the VCSEL axis, from the n-InP spacer layer, through the MQW active region, the p-InP spacer layer, and the bonded interface, to the first GaAs layer of the top DBR. Without interface charges, the small valence band offset of 0.05 eV would only slightly hinder hole injection. In comparison, an interface charge density of $10^{13} \, cm^{-2}$ creates a potential barrier for holes that is more than 0.5 eV high. At lasing threshold, these interface defects create an

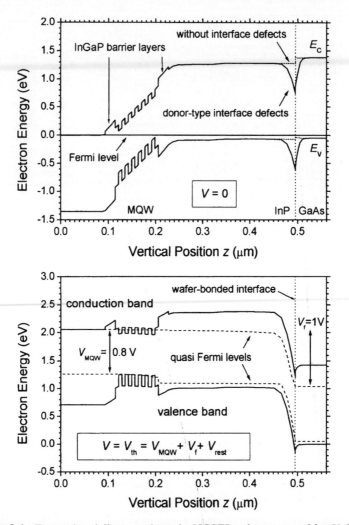

Figure 8.4: Energy band diagram along the VCSEL axis: top, zero bias V; bottom, at threshold voltage V_{th}.

additional voltage drop of $V_f \approx 1$ V, which increases the threshold voltage significantly (Fig. 8.4). Ideally, the threshold voltage should not be much larger than the MQW Fermi level separation of $V_{MQW} \approx 0.8$ V. The fit to the measured current–voltage curve results in an interface defect density of 1.7×10^{13} cm^{-2} for this device [198].

A critical design issue of VCSELs is the reduction of current leakage. In general, two leakage mechanisms can be distinguished near the active region: vertical and lateral leakage. Vertical leakage is caused by carriers that leave the active region in the vertical z direction by thermionic emission. The band diagram in Fig. 8.4 illustrates that the thermionic emission of electrons into the lowly doped p-InP spacer is more likely than the leakage of holes in the opposite direction. Vertical electron leakage can be reduced by incorporating InGaP barrier layers with a slightly larger band gap as shown in Fig. 8.4 [199].

Lateral leakage currents are based on two different mechanisms. First is the lateral diffusion of carriers in the active layer. This ambipolar diffusion process is dominated by the diffusivity of holes. Second is the lateral hole drift within the upper p-InP spacer layer, which is enhanced by the electrical resistance of the lowly doped InP layer underneath. Both lateral leakage currents can be clearly distinguished in Fig. 8.5, which plots the two-dimensional (2D) distribution of

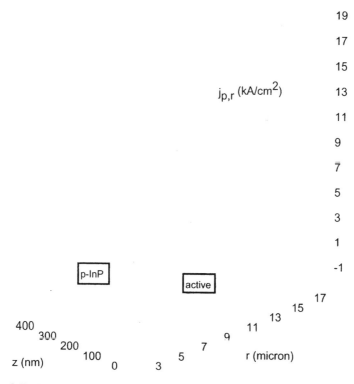

Figure 8.5: Lateral component of the hole current density, $j_{p,r}(r, z)$ ($d_{DBR} = 12\,\mu$m, no oxide layer).

with interface charges   without interface charges

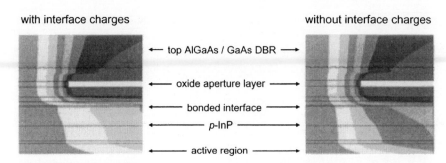

←— top AlGaAs / GaAs DBR —→

←— oxide aperture layer —→

←— bonded interface —→

←——— *p*-InP ———→

←——— active region ———→

Figure 8.6: 2D current contours $j(r, z)$ in the center part of the VCSEL: with (left) and without (right) positively charged defects at the bonded interface (each contour section represents 12.5% of the total current; the left border is the laser axis).

the lateral hole current density, $j_{p,r}(r, z)$. The magnitude of current spreading depends on the hole mobility within the upper spacer layer. The fit to the measured dependence $j_{th}(d_{DBR})$ in Fig. 8.2 gives $\mu_p = 30 \, cm^2/Vs$ for the upper InP layer, which is smaller than expected due to material damage from reactive ion etching.

Oxide confinement within the top DBR further enhances lateral leakage currents. The spreading current is especially severe with positively charged defects at the *p*-side bonded interface. After being funneled through the oxide aperture, holes encounter the high interface resistance, which forces them to spread out laterally before entering the InP layer. This partially eliminates the confinement effect of the oxide layer. The current contour plot in Fig. 8.6 illustrates the influence of charged interface defects. Narrow contours indicate a high current density near the edge of the oxide aperture. Lateral current spreading between oxide layer and bonded interface is much stronger with (left) than without (right) interface defects.

The current crowding at the oxide aperture edge leads to a laterally nonuniform hole distribution in the MQW region, which is most severe in the *p*-side quantum well (Fig. 8.7). This nonuniformity is smoothened as holes move toward the *n*-side quantum wells. The relatively large MQW valence band offset makes it difficult for holes to cross the barriers between the quantum wells. Therefore, the hole concentration is highest in the *p*-side quantum well, as discussed in Section 7.4. Electrons move more easily across the MQW, and they adapt to the nonuniform hole distribution.

8.5   Thermal Analysis

The electrical resistance of interfaces and bulk layers, as well as nonradiative recombination in the active layers are main heat sources in VCSELs (cf. Chap. 6).

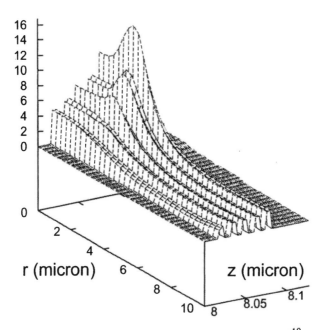

Figure 8.7: 2D surface plot of the hole concentration $p(r, z)$ (10^{18} cm^{-3}) in the MQW active region (5 quantum wells, 2.5 μm oxide aperture radius, no interface charges).

Joule heat depends on the local current density. In the following, aperture size effects on the VCSEL self-heating near threshold are evaluated using a simplified electro-thermal analysis.

At laser threshold, the total heat power is approximately given by the voltage drop across the device times the injection current $V_{th}I_{th}$. The VCSEL threshold voltage V_{th} includes the resistance of various DBR interfaces, which is hard to calculate correctly. Instead, current–voltage measurements on nonoxidized DBR stacks are evaluated [200] to obtain an average p-DBR hole mobility

$$\mu_z(j) = 0.12 \text{ cm}^2/\text{Vs} \times (j \text{ (kA/cm}^2))^{-0.6} \tag{8.1}$$

as a function of the current density j. In the lateral direction, the p-DBR hole mobility $\mu_r = 220$ cm^2/Vs is much larger, and so is the electron mobility in the n-DBR (Table 8.1). The resistance of the p-doped InP/GaAs interface is assumed to be 8×10^{-4} Ω cm^2. A minor heat contribution also arises from the p-contact resistance of 2×10^{-4} Ωcm^2 [200].

The dissipation of heat power inside the device is often characterized by the thermal resistance R_{th}. This number is usually determined as the average active region temperature rise $\Delta T'_{\text{MQW}}$ divided by the heat power. The thermal resistance of VCSELs is typically on the order of 10^3 K/W, one order of magnitude larger than in edge-emitting lasers [201]. This high number results from the smaller size of the heat source as well as from the low thermal conductivity κ_L of DBRs. The small DBR layer thickness restricts the phonon mean free path, and it therefore reduces the thermal conductivity. For our binary AlAs/GaAs n-DBRs, the average thermal conductivity is measured to be about half of the expected average bulk value [202]. The DBR's thermal conductivity is larger in the lateral direction than in the vertical direction [203]. AlGaAs exhibits a strong alloy scattering of phonons. In agreement with these results, anisotropic thermal conductivities are used as given in Table 8.1. An average temperature dependence of $\kappa_L \propto T^{-1.375}$ is included [69], and a value of $\kappa_L = 0.2$ W/cmK is assumed for the oxide layer.

Lateral oxidation allows for very small active areas and very low threshold currents I_{th}. In oxidized 1.55-μm VCSELs, $I_{\text{th}} = 0.8$ mA was reported with an oxide aperture of $d_{\text{ox}} \approx 2.4\,\mu$m, resulting in a threshold current density of $j_{\text{th}} \approx 18\,\text{kA}/\text{cm}^2$ [204]. This large current density is related to lateral carrier leakage and optical scattering losses that increase with smaller oxide aperture. The measured dependence $j_{\text{th}}(d_{\text{ox}})$ is given in Fig. 8.8 (dots), and it is compared to the ideal case of constant threshold current density $j_{\text{th}} = 1\,\text{kA}/\text{cm}^2$ (dashed line).

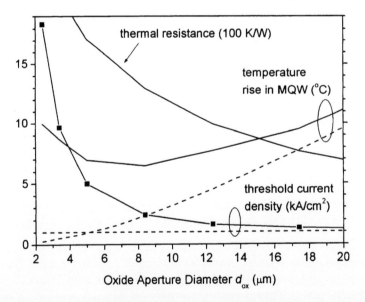

Figure 8.8: Threshold current density j_{th}, thermal resistance R_{th}, and active region heating ΔT_{MQW} as a function of the oxide aperture diameter d_{ox} (dots, measurement; lines, simulation; dashed lines represent the case of constant threshold current density).

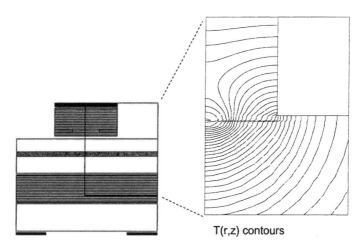

Figure 8.9: Temperature distribution $T(r, z)$ as calculated in the center part of the VCSEL.

Figure 8.9 illustrates the calculated temperature distribution $T(r, z)$ in the center part of the device. The heat flows perpendicular to the isothermal lines and it uses the top DBR for lateral dissipation. Thus, even with a bottom heat sink, the thermal conductivity of layers above the heat source matters. Downsizing the active area causes a dramatic increase of the thermal resistance (Fig. 8.8). This is attributed to the reduced initial escape area for the heat generated in the center of the device. In our case, the simulations result in $R_{th} \propto d_{ox}^{-0.68}$.

The calculated average temperature rise ΔT_{MQW} inside the active region of the oxidized VCSEL is plotted in Fig. 8.8. With constant threshold current density (dashed line), the self-heating at threshold vanishes for very small apertures. This is despite the strong increase of R_{th} since the heat power is reduced even stronger ($I_{th} \propto d_{ox}^2$). In the measured case of increasing threshold current density (solid line), a minimum temperature rise occurs at $d_{ox} = 8\,\mu\text{m}$, which is in good agreement with the experimental results [204]. This optimum aperture size gives the lowest heat power generation. Self-heating can be further minimized by reducing lateral leakage and optical losses with low apertures.

8.6 Optical Simulation

Calculation of the internal optical field is one of the most challenging tasks of VCSEL simulations, especially when oxide confinement layers are involved [205]. A precise optical analysis requires elaborate solutions of Maxwell's equations for an open resonator [206, 207, 208]. Simplified optical models are often applied in practical VCSEL simulations.

For our nonoxidized VCSELs, we here describe the combination of a simple optical VCSEL model with experimental results in order to correctly simulate the VCSEL performance near threshold [167]. The model is based on the effective index method [209], and it decouples vertical and lateral mode profiles. The lateral mode shape is approximated by Bessel functions, which are solutions to the Helmholtz equation for cylindrical waveguides (Fig. 8.10).

The transmission matrix method is utilized to obtain the standing optical wave in the vertical direction (Fig. 8.11). DBR layers are quarter-wavelength thick and the standing wave exhibits antinodes (peaks) or nodes (nulls) at the mirror interfaces. Nodes are located at the bonded interfaces so that interface absorption is minimized. The resonance wavelength λ_{cav} is given by the optical mirror distance (index $\times$ thickness). For maximum gain, the MQW region is located at the peak of the standing wave. Table 8.1 lists the material parameters n_r and α_0 essential for the optical simulation. These parameters shall now be discussed in more detail.

The temperature dependence of the refractive index $n_r(T)$ has considerable impact not only on the emission wavelength but also on DBR reflectivity, slope efficiency, and threshold gain [210]. The value of dn_r/dT is different for each material (cf. Table 4.3) so that the interface refractive index step changes

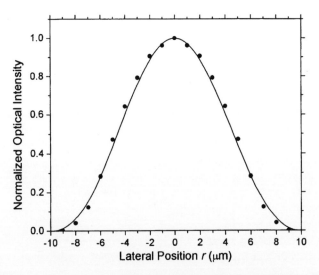

Figure 8.10: Lateral optical intensity profile for the fundamental mode (dots, near-field measurement; line, squared zero-order Bessel function).

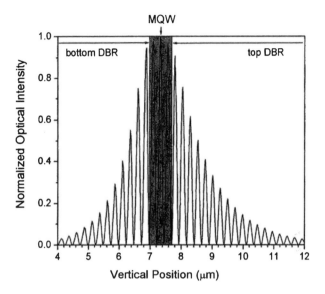

Figure 8.11: Vertical optical intensity profile.

with temperature. For realistic continuous-wave simulations, the refractive index of each layer is determined as a function of the local temperature. Thermal expansion is also included but its influence is negligible.

At room temperature, the measured emission wavelength is $\lambda_{cav} = 1542$ nm. In order to reproduce this number in the simulation, the InP spacer thickness is slightly adjusted. Inhomogeneities of several tens of nanometers across the InP wafer are not surprising. Slight deviations of other layers or of refractive indices are also possible but hard to separate.

Photon absorption in our VCSELs is mainly caused by holes (cf. Section 7.2.3). However, scattering losses at the rough sidewall of the top DBR pillar must be added to the average bulk absorption of $8\,\text{cm}^{-1}$. These sidewall losses depend on the pillar diameter d_{DBR}. They are hard to calculate but they can be extracted from the measured slope efficiency [211]. The fit to the experimental characteristic $\eta_d(d_{DBR})$ in Fig. 8.2 gives the p-DBR optical loss as

$$\alpha_{p-DBR} = 8\,\text{cm}^{-1} + \frac{2.39 \times 10^{-5}\text{cm}}{d_{DBR}^2}. \tag{8.2}$$

This function leads to $\alpha_{p-DBR} \approx 25/\text{cm}$ for $d_{DBR} = 12\,\mu\text{m}$ as given in Table 8.1. Such a combination with measurements helps to find realistic numbers for model parameters that are unknown and hard to calculate.

8.7   Temperature Effects on the Optical Gain

In vertical-cavity lasers, the wavelength λ_{gain} of maximum gain needs to be adjusted to the emission wavelength λ_{cav}. The gain offset $\lambda_{gain} - \lambda_{cav}$ depends on the temperature since λ_{cav} shifts with changing refractive index and λ_{gain} shifts with changing MQW band gap (Fig. 8.12). The adjustment of the gain offset is a critical VCSEL design issue in order to reduce temperature effects on the threshold current I_{th} [212]. Minimum temperature sensitivity ($dI_{th}/dT = 0$) is obtained at the minimum of the threshold current temperature characteristic $I_{th}(T)$. It is a common VCSEL design rule that the threshold current $I_{th}(T)$ has its minimum at zero gain offset [213, 214]. However, this design rule does not seem to apply to InP-based long-wavelength VCSELs for which the lowest threshold current is measured at considerable negative gain offset (Fig. 8.12) [194, 195, 215]. We shall investigate this phenomenon in the following.

In our VCSEL, the emission wavelength shift with temperature is $d\lambda_{cav}/dT = 0.11$ nm/K. The photoluminescence (PL) peak wavelength shift with temperature was measured on similar MQWs to be $d\lambda_{PL}/dT = 0.54$ nm/K, giving a thermal band gap shrinkage of $dE_g/dT = -0.28$ meV/K [216], which is about two-thirds

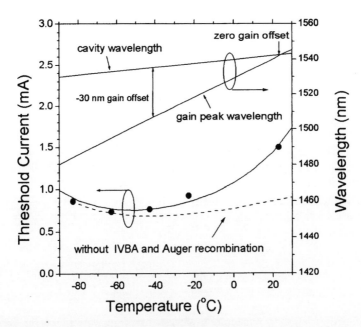

Figure 8.12: Wavelength shifts and threshold current as a function of temperature (pulsed operation; dots, measurement; lines, simulation).

of typical GaAs values [69]. The PL peak is only a few nanometers blue-shifted from the gain peak (cf. Fig. 5.11). Zero gain offset is at 24°C.

The calculated threshold current density $j_{th}(T)$ agrees very well with the experimental data (Fig. 8.12). IVBA and Auger recombination have been identified as the most temperature-sensitive loss mechanisms in 1.55-μm VCSELs [167]. Exclusion of those two loss mechanisms from the model gives the dashed curve in Fig. 8.12. As result of this exclusion, the position of the $j_{th}(T)$ minimum is only slightly shifted to −40°C, indicating that the minimum of the threshold current is a gain-related effect.

Based on the good agreement of our model with experimental data, temperature effects on the calculated gain are now evaluated in more detail. The effect of the temperature on the MQW gain spectrum is plotted in Fig. 8.13 for constant carrier concentration. The dots indicate the VCSEL emission wavelength $\lambda_{cav}(T)$ and the modal gain at each temperature. At 20°C, the gain peak wavelength $\lambda_{gain} = 1540$ nm is only 2 nm smaller than the emission wavelength λ_{cav}. As the temperature is reduced, the gain peak is not only blue-shifted, the maximum

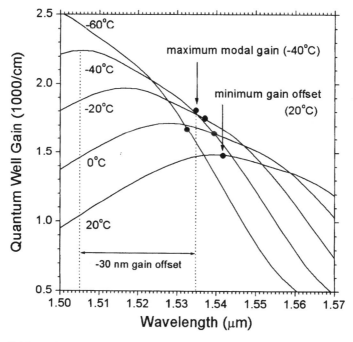

Figure 8.13: MQW gain spectra at constant carrier concentration for different temperatures, T. The dots give the emission wavelength $\lambda_{cav}(T)$ and the modal gain.

of the gain spectrum also increases strongly ($dg_{max}/dT = -10\,\mathrm{cm}^{-1}\,\mathrm{K}^{-1}$). This increase is attributed to the decreased Fermi spreading of carriers in the quantum wells. The maximum modal gain is reached at $-40°\mathrm{C}$ corresponding to $-30\,\mathrm{nm}$ gain offset. Without loss effects and with almost constant threshold gain, this maximum modal gain results in a minimum threshold current at the same temperature (dashed curve in Fig. 8.12). Thus, the simulation clearly shows that about $-30\,\mathrm{nm}$ gain offset is necessary in these 1.55-μm VCSELs to obtain minimum threshold current. The optical cavity mode does not receive maximum gain at zero gain offset, as commonly assumed.

Further Reading

- T. E. Sale, *Vertical-Cavity Surface Emitting Lasers*, Wiley, New York, 1995.
- C. Wilmsen, H. Temkin, and L. A. Coldren (Eds.), *Vertical-Cavity Surface Emitting Lasers*, Cambridge Univ. Press, Cambridge, UK, 1999.

Chapter 9

Nitride Light Emitters

This chapter presents the simulation of milestone device structures of blue light-emitting diodes and laser diodes fabricated by Nakamura et al. on GaN. Performance-limiting physical mechanisms like self-heating, current crowding, and carrier leakage are investigated. Unique material properties of wurtzite nitride compounds such as high-field electron mobility and built-in polarization fields are discussed.

9.1 Introduction

Light-emitting diodes (LEDs) and laser diodes based on GaN were pioneered by Shuji Nakamura and coworkers in the 1990s, and they soon became commercially available [217]. The active region of these devices is made of InGaN, which can cover the wide wavelength range from violet to red (Fig. 9.1). At room temperature, GaN has a band gap wavelength of $\lambda_g = 363$ nm and InN exhibits $\lambda_g = 657$ nm. Especially blue and green nitride LEDs are now widely used, for instance, in full-color displays and in traffic signals. Nitride laser diodes have great potential in a number of applications such as optical data storage, printing, full-color displays, chemical sensors, and medical applications. The data capacity of digital versatile disks (DVDs) can be increased from 4.7 to more than 15 Gbytes by using blue InGaN lasers instead of red AlInGaP lasers.

Nitride devices are typically grown in the hexagonal (wurtzite) crystal system, which is energetically favored over cubic (zinc blende) nitrides (cf. Fig. 2.10). Band gaps and lattice constants for wurtzite nitrides are given in Fig. 9.1. Lattice-matched substrates are currently not available for GaN. Sapphire or SiC are mainly used as substrates with lattice constants a_0 of 4.758 and 3.08Å, respectively. Despite intense research efforts worldwide on nitride devices, major technological challenges, as well as the need for a more detailed understanding of microscopic physical processes, remain. Numerical simulation can help to investigate those processes and to establish quantitative links between material properties and measured device performance. Many general nitride material parameters are given in Part I, for instance in Table 2.7. Additional and more unique nitride properties are discussed in the next Section, including alloy bowing parameters. In Section 9.3, we present the simulation and analysis of nitride LEDs using APSYS.

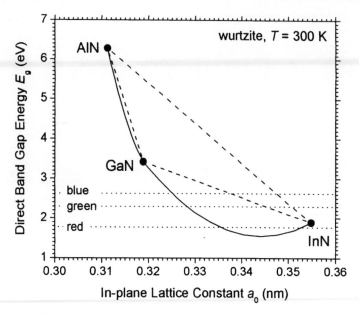

Figure 9.1: Band gaps and lattice constants for hexagonal nitride compounds. The solid lines give the average of measured band gap bowing for ternary alloys AlGaN and InGaN, respectively.

Section 9.4 investigates the physics and performance of blue laser diodes using LASTIP.

9.2 Nitride Material Properties

9.2.1 Carrier Transport

A major handicap with nitride semiconductors is the high activation energy of the common magnesium acceptor, which causes the hole concentration to be considerably smaller than the acceptor concentration. A magnesium activation energy of 170 meV was reported for GaN [218], which is assumed to increase by 3 meV per percent of Al for AlGaN. In n-type material, the silicon donor activation energy is about 20 meV [219].

GaN room-temperature electron mobilities near 900 cm^2/Vs have been measured with low doping, which are reduced by more than an order of magnitude with high doping (Fig. 9.2) [220]. Monte Carlo simulations of various nitride alloys have recently been used to establish analytical expressions for the electron mobility as a function of donor concentration N_D, temperature T, and electrostatic field

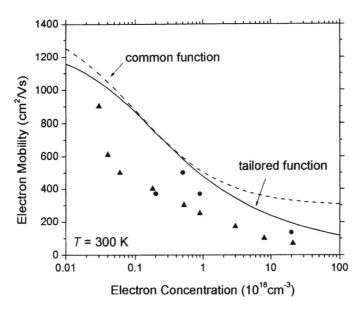

Figure 9.2: Doping effects on the GaN electron mobility — comparison of measurements (triangles [220], circles [219]) and simulations (solid line, Eq. (9.1) [221]; dashed line, Eq. (3.28) [63]).

F [63]. The low-field mobility $\mu_n(N_D, T)$ can be approximated by the common functions given in Section 3.6 with the nitride parameters of Table 3.2. A more tailored function for n-doped GaN is given in [221] as

$$
\frac{1}{\mu_n^{\text{GaN}}(N_D, T)} = A \frac{N_D}{N_0} \left(\frac{T}{T_0}\right)^{-3/2} \ln\left[1 + 3\left(\frac{T}{T_0}\right)^2 \left(\frac{N_D}{N_0}\right)^{-2/3}\right]
$$
$$
+ B \left(\frac{T}{T_0}\right)^{3/2} + \frac{C}{\exp[T_1/T] - 1} \tag{9.1}
$$

with the parameters $A = 2.61\,\text{Vs/m}^2$, $B = 2.9\,\text{Vs/m}^2$, $C = 170\,\text{Vs/m}^2$, $N_0 = 10^{17}\,\text{cm}^{-3}$, $T_0 = 300\,\text{K}$, and $T_1 = 1065\,\text{K}$. Figure 9.2 compares both fit functions with measured low-field electron mobilities in GaN. For high-field mobilities, the unique fit function

$$
\mu_n(F) = \frac{\mu_0 + \frac{v_s}{F_c}\left(\frac{F}{F_c}\right)^{n_1 - 1}}{1 + A_n \left(\frac{F}{F_c}\right)^{n_2} + \left(\frac{F}{F_c}\right)^{n_1}} \tag{9.2}
$$

was derived with the parameters given in Table 9.1. The low-field mobility μ_o is listed in Table 3.2. Results are plotted in Fig. 9.3 for different alloys.

Hole mobilities are much less investigated. For Mg-doped GaN, $\mu_p = 15 \text{ cm}^2/\text{Vs}$ has been measured with low hole concentration and $2 \text{ cm}^2/\text{Vs}$ with high hole concentration [222]. This results in a hole conductivity that hardly varies with the doping density. Mobility reduction with temperature elevation was measured

Table 9.1: Parameters for the High-Field Electron Mobility Function Given in Eq. (9.2) [63]

Symbol Unit	v_s (10^7cm/s)	F_c (kV/cm)	n_1 —	n_2 —	A_n —
GaN	1.91	221	7.20	0.786	6.20
InN	1.36	52	3.85	0.608	2.26
AlN	2.17	447	17.37	0.855	8.73
$Al_{0.2}Ga_{0.8}N$	1.12	366	5.32	1.04	3.23
$In_{0.2}Ga_{0.8}N$	1.04	208	4.72	1.02	3.62

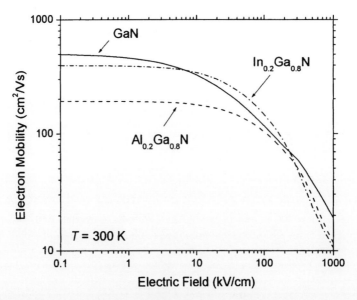

Figure 9.3: Electron mobility as a function of the electrostatic field for bulk nitride alloys with 10^{18} cm^{-3} doping density [63].

as $\mu_n \propto T^{-1.5}$ [219] and $\mu_p \propto T^{-6}$ [223]. This temperature effect on the hole mobility is much stronger than with other III–V compounds [62].

Defect-related recombination is known to be the main carrier loss mechanism in nitride devices. The Shockley–Read–Hall recombination lifetime of electrons and holes is on the order of 1 ns; however, the type and concentration of recombination centers are sensitive to the technological process. For bulk GaN layers, a spontaneous emission parameter B of 1.1×10^{-8} cm^3 s^{-1} was measured [224], which is almost three orders of magnitude higher than theoretical estimations (2×10^{-11} cm^3 s^{-1} [225]). The spontaneous recombination rate in quantum wells is larger than in bulk layers, and it can be calculated by integration of the spontaneous emission spectrum. A small GaN Auger parameter of $C = 10^{-34}$ cm^6 s^{-1} is estimated using its band gap dependence in other materials (Fig. 3.12). Thus, even with large carrier concentrations, Auger recombination in nitride materials is negligible.

9.2.2 Energy Bands

A wide spectrum of band gap bowing parameters has been obtained for ternary nitride alloys due to differences in growth and measurement conditions [36, 226]. For unstrained layers with a low mole fraction of the alloy element ($x < 0.3$), we employ the following room-temperature relations for the direct band gap (eV)

$$E_g(\text{In}_x\text{Ga}_{1-x}\text{N}) = 1.89x + 3.42(1 - x) - 3.8x(1 - x) \qquad (9.3)$$

$$E_g(\text{Al}_x\text{Ga}_{1-x}\text{N}) = 6.28x + 3.42(1 - x) - 1.3x(1 - x), \qquad (9.4)$$

which is plotted in Fig. 9.1. A band offset ratio of $\Delta E_c / \Delta E_v = 0.7/0.3$ is assumed for InGaN/GaN as well as for AlGaN/GaN, which corresponds to an average of reported values for each case [33, 227].

For the strained InGaN quantum wells investigated in this chapter, the conduction bands are assumed to be parabolic and the nonparabolic valence bands are computed by the 6×6 $\vec{k} \cdot \vec{p}$ method (cf. Section 2.2.4). Binary effective mass parameters, lattice constants, and elastic constants from Table 2.7 are linearly interpolated to obtain InGaN values. GaN values are used for the deformation potentials [29]. Calculated valence bands are shown in Fig. 9.4. Similar band structures have been calculated by other authors [228]. For such nitride layers, grown along the c axis, the quantum well strain is biaxial, lowering the crystal-field split-hole (CH) band but hardly separating heavy-hole (HH) and light-hole (LH) bands; i.e., strain is not as effective in reducing the threshold carrier concentration as in GaAs- or InP-based lasers. Uniaxial strain promises lower threshold currents [229]. Thermal band gap shrinkage is considered with $dE_g/dT = -0.6$ meV/K [32]. Band gap renormalization caused by carrier–carrier interaction was found to depend on the

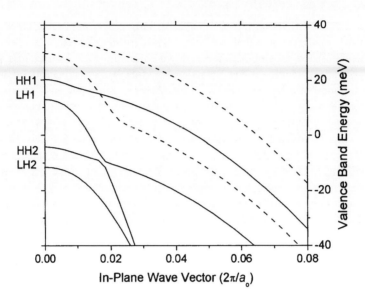

Figure 9.4: Valence band structure of a 4-nm $In_{0.15}Ga_{0.85}N/In_{0.02}Ga_{0.98}N$ quantum well (HH, heavy-hole bands; LH, light-hole bands; solid lines, 1.6% compressive strain; dashed lines, upper two bands without strain).

quantum well 2D carrier concentration as

$$\Delta E_g = -\zeta N_{2D}^{1/3} \tag{9.5}$$

with $\zeta = 6 \times 10^{-6}$ eV cm$^{2/3}$ being independent of well thickness and alloy composition [230].

9.2.3 Polarization

Built-in electrical fields in semiconductors can be caused by spontaneous polarization P_{sp} or by strain-induced polarization P_{piezo}. These polarization effects were found to be much stronger in c-face nitrides than in other III–V compounds [231]. With the growth direction along the c axis, both types of polarization add up to the surface charge density

$$q\sigma_{2D} = P_{sp} + P_{piezo} = P_{sp} + 2\frac{a_s - a_0}{a_0}\left(e_{31} - \frac{C_{13}}{C_{33}}e_{33}\right) \tag{9.6}$$

with the piezoelectric constants e_{33} and e_{31}, the elastic constants C_{13} and C_{33}, and the wurtzite lattice constant a_0 (a_s is the substrate value). The material parameters are given in Table 2.7 and in Table 9.2 for binary compounds. For ternary

Table 9.2: Polarization Parameters for Nitride Materials [232]

Parameter	Symbol	Unit	InN	GaN	AlN
Spontaneous polarization	P_{sp}	$C\,m^{-2}$	-0.042	-0.034	-0.09
Piezoelectric constant (z)	e_{33}	$C\,m^{-2}$	0.810	0.670	1.50
Piezoelectric constant (x, y)	e_{31}	$C\,m^{-2}$	-0.410	-0.340	-0.53

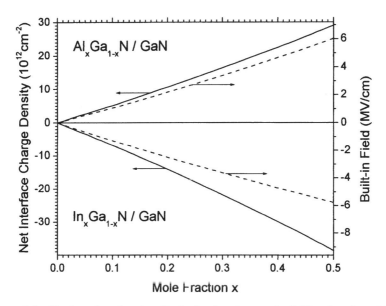

Figure 9.5: Net interface density of polarization charges (solid lines) and resulting electrostatic field (dashed lines) for ternary alloys grown on GaN.

compounds, these numbers may be interpolated linearly; however, some nonlinear behavior has recently been observed [233]. Since different alloy compositions show different polarization, net charges remain at their interfaces. For ternary nitride alloys grown on GaN, Fig. 9.5 plots the net interface charge density as well as the resulting electrostatic field within the alloy. These built-in fields significantly affect the properties of quantum wells [234]. The corresponding band diagram is illustrated in Fig. 9.6. The wider the quantum well, the more separated the electrons and holes, and the smaller the optical gain and spontaneous emission. Similar to the electroabsorption effect illustrated in Fig. 10.1, the transition energy is reduced by the built-in field, leading to a red-shift of the emission wavelength. However, with increasing carrier injection into the quantum well, charge

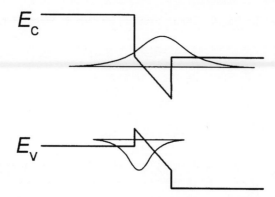

Figure 9.6: Band edges of conduction (E_c) and valence bands (E_v), confined energy levels, and wave functions for a 4-nm-thick $In_{0.13}Ga_{0.87}N$/GaN quantum well with 1 MV/cm polarization field.

screening is expected to reduce polarization field effects [235]. These screening effects result in a blue-shift of the emission wavelength as measured on nitride LEDs [236].

9.2.4 Refractive Index

Optical reflection and waveguiding mainly depend on the refractive index profile inside the device. For photon energies close to the band gap, the refractive index is a strong function of wavelength. For binary nitride compounds, the modeling of this function is discussed in Section 4.2.2. In ternary alloys, those models need to be modified to include the composition as parameter. For the design of optical waveguides, the compositional change of the refractive index is often more important than its absolute value. In comparison to refractive index measurements on $Al_xGa_{1-x}N$ ($x < 0.38$), the Adachi model (Eq. (4.34)) was found to be most accurate using the parameters [120]

$$A(x) = 9.827 - 8.216x - 31.59x^2 \qquad (9.7)$$

$$B(x) = 2.736 + 0.842x - 6.293x^2. \qquad (9.8)$$

Reliable refractive index measurements on $In_xGa_{1-x}N$ are currently not available, so a linear interpolation between binary parameters A and B is chosen here (cf. Table 4.2)

$$A(x) = 9.827(1 - x) - 53.57x \qquad (9.9)$$

$$B(x) = 2.736(1 - x) - 9.19x. \qquad (9.10)$$

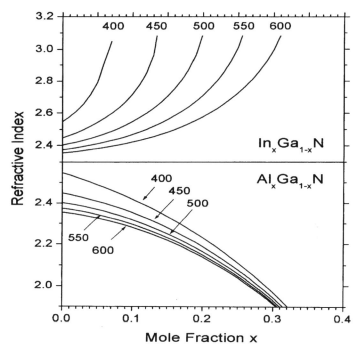

Figure 9.7: Refractive index for different alloy compositions with the wavelength given as a parameter (nm).

In addition we employ the band gap relations given in Eqs. (9.3) and (9.4), respectively. The resulting refractive index is plotted in Fig. 9.7. Other authors employ Sellmeier-type models for GaN (Eq. (4.28)) and predict the ternary refractive index by adjusting the photon energy according to the band gap change [237]. For photon energies close to the band gap, this phenomenological approach is less reliable than the Adachi model (cf. Section 4.2.2). In InGaN quantum wells, the refractive index is modified due to the increased transition energy and other effects [238].

9.2.5 Thermal Conductivity

Measurements of the thermal conductivity are hardly published for nitride alloys. For GaN, a room-temperature value of 1.3 W/cmK is reported in [38]. Further thermal parameters for nitride binaries are given in Table 6.1. Alloy scattering can reduce the thermal conductivity of semiconductor compounds by more than one order of magnitude (cf. Fig. 6.1). However, the bowing parameter C_{ABC} in Eq. (6.7) is unknown for ternary nitride alloys. In addition, phonon scattering at the interfaces of thin layers is known to reduce their thermal conductivity

Table 9.3: Layer Structure and Room-Temperature Parameters of the InGaN/GaN LED

Parameter Unit	d (nm)	N_{dop} (1/cm^3)	μ (cm^2/Vs)	n_r —	κ_L (W/cmK)
p-GaN	500	1×10^{20}	10	2.45	1.3
p-Al$_{0.3}$Ga$_{0.7}$N	100	2×10^{19}	5	1.94	0.4
In$_{0.2}$Ga$_{0.8}$N	2	—	200	3.07	0.1
n-In$_{0.02}$Ga$_{0.98}$N	50	1×10^{18}	570	2.48	0.9
n-Al$_{0.3}$Ga$_{0.7}$N	100	3×10^{18}	300	1.94	0.4
n-GaN	4000	3×10^{18}	410	2.45	1.3

Note. d, layer thickness; N_{dop}, doping density; μ, majority carrier mobility (low field); n_r, refractive index (wavelength 450 nm); κ_L, thermal conductivity.

considerably [202]. Assuming significant impact of both these effects, the alloy thermal conductivities used in this chapter are estimated and given in Tables 9.3 and 9.4.

9.3   InGaN/GaN Light-Emitting Diode

9.3.1   Device Structure

As a practical example, we here investigate single-quantum-well (SQW) blue light-emitting diodes (LEDs) as described in [239]. These devices exhibit an output power of 4.8 mW at 20 mA injection current and 3.1 V forward voltage. The low voltage results in a relatively high power efficiency of 7.7% (ratio of light power to electrical power). The external quantum efficiency is as high as 8.7% despite the large dislocation density of about 10^{10} cm^{-2} [240]. Fluctuations of the Indium composition within the InGaN quantum well are assumed to localize carriers in radiative centers and to prevent them from recombining nonradiatively at dislocations [241].

The device was grown by metal organic chemical vapor deposition (MOCVD) on c-face sapphire. The layer sequence is given in Table 9.3 and the band diagram is shown in Fig. 9.8. The 2-nm-thick undoped quantum well is sandwiched between n-In$_{0.02}$Ga$_{0.98}$N and p-Al$_{0.3}$Ga$_{0.7}$N. After growth, the structure was partially etched to expose the n-GaN layer. A Ti/Al contact was deposited on the n-GaN layer and a semitransparent Ni/Au contact on the top p-GaN layer (Fig. 9.9). Part of the top p-contact was covered by a nontransparent bonding pad. The wafer was cut into rectangular LED chips (300 × 300 μm^2), which were molded onto a lead frame.

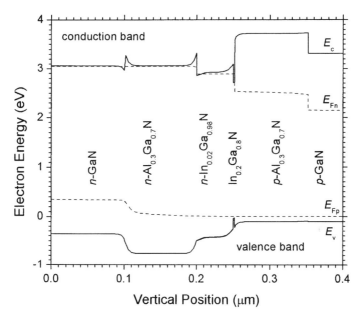

Figure 9.8: Vertical energy band diagram of the LED active region at 3.1 V forward bias (E_c, E_v, band edge energies; E_{Fn}, E_{Fp}, quasi-Fermi levels; without polarization charges).

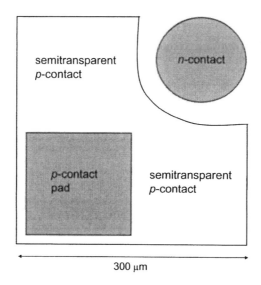

Figure 9.9: Schematic top view of the LED.

9.3.2 Polarization Effects

Recombination mechanisms in real InGaN quantum wells are still not fully understood. Polarization, nonuniform indium distribution, and exciton effects may have a major influence [242]. We start here with a simple free-carrier model to calculate the spontaneous emission spectrum for the nonsymmetric quantum well. According to experimental results [236], the interface polarization charges derived from Eq. (9.6) are assumed to be partially screened by charged defects so that only about half of them contribute to the built-in polarization field. The effect of these net polarization charges on the LED quantum well is shown in Fig. 9.10, including screening by electrons and holes. Without polarization, the internal electrical field of the pn-junction is positive and the holes accumulate at the p-side of the quantum well. With polarization, the electrical field is strongly negative within the quantum well and the holes accumulate at the n-side. In both cases, the peak of the electron distribution is near the center of the quantum well. Due to the more nonsymmetric quantum barriers, there are less confined electrons without polarization. Built-in polarization allows for more quantum levels.

Calculated spectra of the spontaneous emission in Fig. 9.11 exhibit a reduction of the peak emission intensity due to the polarization field. We assume here uniform current injection at room temperature. Lorentzian broadening with an intraband scattering time of 0.1 ps gives a full-width at half-maximum (FWHM) of 8 nm without polarization. Due to the additional confined quantum states, built-in polarization results in a broader line shape with FWHM = 17 nm. The measured line broadening of 25 nm [239] also includes the effect of fluctuations in quantum well composition, width, strain, and polarization, which are not considered in our model [243]. Built-in polarization red-shifts the emission peak by only 1 nm due to the opposite tilt of the quantum well in both cases (Fig. 9.10). With polarization, the integrated spontaneous emission rate is enhanced by about 20%.

9.3.3 Current Crowding

The three-dimensional (3D) structure of the LED is approximated in our 2D APSYS simulation as shown in Fig. 9.12. The pn-junction is 160 μm wide and 300 μm long, giving about the same top emission area as with the real device (Fig. 9.9). The simulated device is uniform in longitudinal (z) direction so that all xy cross-sections are identical. The Ni/Au semitransparent p-contact is treated as a perfect conductor and the current is uniformly injected across the top surface. However, due to the low conductivity of p-doped nitrides, strong current crowding is observed in the simulation (Fig. 9.12). The carriers choose the current path of lowest total resistance, which depends on the conductivity of each material involved.

In our case of a perfect p-contact, most of the radiation is generated close to the inner edge of the device. Figure 9.13 plots the 2D distribution of the emission

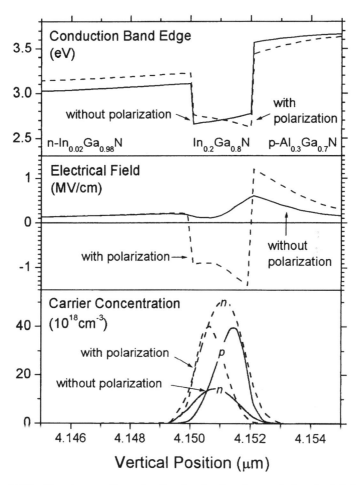

Figure 9.10: Quantum well conduction band edge, internal electrical field, and carrier profiles with (dashed lines) and without (solid lines) built-in polarization at 3.1 V bias and 200 A/cm^2 current density ($T = 300$ K).

rate near the edge. Due to the polarization field, the holes are localized at the lower edge of the quantum well and cause the radiation peak there. At the simulated high current injection ($I = 100$ mA), holes also leak out into the InGaN barrier region, causing emission at shorter wavelengths as observed experimentally [244].

The leakage of carriers from the quantum well is supported by the self heating of the device. An external thermal resistance of $R_{th} = 200$ K/W is assumed in the simulation, which represents the sapphire substrate as well as the packaging [245]. The internal temperature distribution is shown in Fig. 9.14. As expected, the maximum

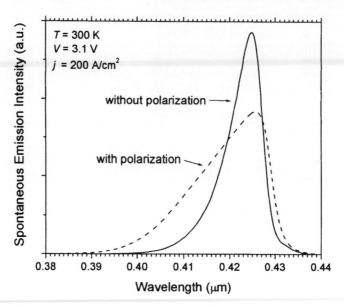

Figure 9.11: Spontaneous emission spectrum for the nonsymmetric quantum well showing the effect of built-in polarization.

temperature rise of $\Delta T_{\max} = 73$ K occurs at the edge of the device. The calculated self-heating is in good agreement with experimental observations on similar devices [245]. However, due to the relatively high thermal conductivity of GaN, the internal LED temperature is almost uniform. It is 97.5°C at the GaN/sapphire interface, corresponding to a 72.5-K temperature rise above the 25°C ambient temperature. In other words, the self-heating is due to the high thermal resistance of substrate and packaging. Current crowding and self-heating cause a further red-shift of the emission peak to 435 nm wavelength. The measured emission wavelength of 450 nm is slightly longer, which indicates a nonuniform indium distribution along the quantum well.

9.3.4 Quantum Efficiency

The external quantum efficiency η_{ext} gives the average number of photons emitted per injected electron. It can be separated into internal photon generation efficiency and optical photon extraction efficiency

$$\eta_{\text{ext}} = \eta_{\text{int}} \times \eta_{\text{opt}}. \qquad (9.11)$$

These absolute efficiencies should not be confused with the differential efficiencies of laser diodes (cf. Chap. 7). We shall first discuss the internal efficiency, η_{int}.

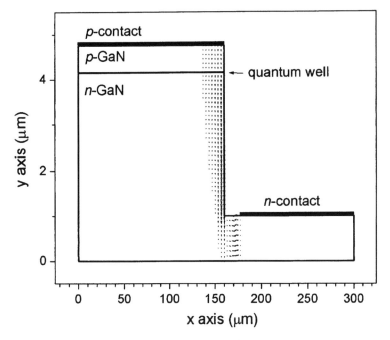

Figure 9.12: Vector plot of the current distribution in our 2D LED model (total current = 100 mA; the size of the arrows is scaled with the current density and no arrows are shown for less than 10% of the maximum current density).

Figure 9.15 plots the calculated internal light power $P_{int}(I)$ as well as $\eta_{int}(I)$. The internal efficiency is the fraction of photons generated per injected electron. Without any carrier losses, the maximum possible value is $\eta_{int} = 1$. This efficiency is directly related to the $P_{int}(I)$ curve

$$\eta_{int} = \frac{q}{h\nu} \frac{P_{int}}{I} = \frac{I_{spon}}{I_{spon} + I_{nr} + I_{leak}} \tag{9.12}$$

($h\nu$, photon energy). It can also be calculated from the ratio of the spontaneous recombination current I_{spon} to the total current. The total current includes contributions to nonradiative recombination within the quantum well (I_{nr}) and carrier leakage from the quantum well (I_{leak}). Leaking carriers eventually recombine outside the quantum well, where we assume nonradiative lifetimes of 0.1 ns for AlGaN and 1 ns for all other layers. The nonradiative carrier lifetime τ_{nr} inside the InGaN quantum well is considered large but it is hard to measure. In our simulation, we assume the ideal case of an extremely long lifetime $\tau_{nr} = 1\mu s$, which

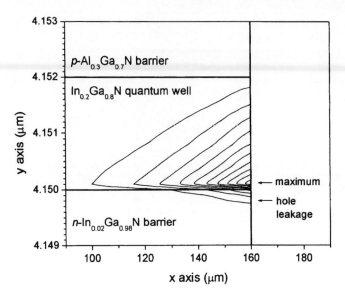

Figure 9.13: Contour plot of the emission rate near the edge of the LED at 100 mA total current (resolution limited by mesh size).

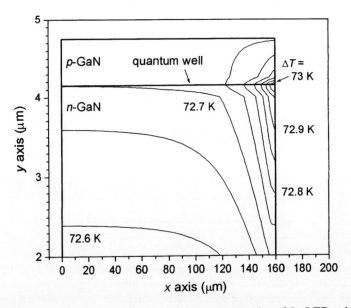

Figure 9.14: Temperature distribution $T(x, y)$ near the edge of the LED at 100 mA total current.

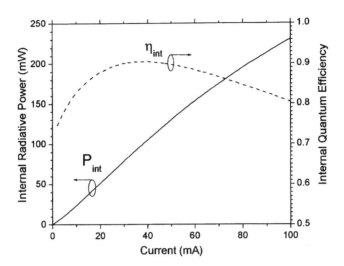

Figure 9.15: Internal light power (solid) and internal quantum efficiency (dashed) as a function of current.

renders nonradiative recombination irrelevant inside the quantum well ($I_{nr} = 0$). Thus, our calculated maximum of $\eta_{int} = 0.91$ is limited by carrier leakage. With higher current and higher temperature, the leakage losses increase and reduce the efficiency in Fig. 9.15.

After photons are generated inside the semiconductor, they need to be extracted from the LED to be visible. The optical extraction efficiency η_{opt} gives the fraction of photons that escape from the device per photon generated inside. Photon escape is mainly limited by total internal reflection. Our device is encapsulated in an epoxy dome with a refractive index of about $n_e = 1.5$. This gives a critical angle of 37° (cf. Eq. (4.56)), which is considerably wider than that of 26° in conventional LEDs. By integration over the escape cone, the optical quantum efficiency of the top surface can be approximated as

$$\eta_{opt}^{top} = \frac{1}{2} \left(1 - \sqrt{1 - \frac{n_e}{n_s}} \right) \left(1 - \frac{(n_e - 1)^2}{(n_e + 1)^2} \right). \qquad (9.13)$$

The last factor accounts for the transmission from the epoxy to air. With a semiconductor index of about $n_s = 2.5$, this equation gives $\eta_{opt}^{top} = 0.096$ for our top surface. Multiple reflections inside the device and emission from other surfaces shall give a larger value for the total optical efficiency. Ray tracing based on simple plane wave theory is often used to simulate the photon extraction from LEDs (cf. Chap. 4); however, the 2D ray tracing model currently included in APSYS is not able to give realistic results for our 3D device. For low photon absorption, an upper limit of $\eta_{opt} = 0.68$ was estimated for blue LEDs on a sapphire substrate [246].

9.4   InGaN/GaN Laser Diode

The LASTIP laser model self-consistently combines wurtzite band structure and gain calculations with 2D simulations of wave guiding, carrier transport, and heat flux. As a practical nitride laser example, we investigate here ridge–waveguide Fabry–Perot laser diodes with two InGaN/GaN quantum wells as demonstrated in [247]. These devices exhibit high-power lasing up to 420 mW in continuous-wave operation.

9.4.1   Device Structure

The layer structure is given in Table 9.4 and the band diagram in Fig. 9.16. The active region comprises two 4-nm-thick InGaN quantum wells with 1.6% compressive strain as well as an AlGaN electron stopper layer. It is sandwiched between GaN separate confinement layers and superlattice (SL) AlGaN/GaN cladding layers. The thickness of the n-side cladding layer was increased from 600 to 1200 nm to reduce the penetration of the laser light into the GaN substrate. In the simulation, we replace the superlattice by bulk $Al_{0.07}Ga_{0.93}N$. The vertical profiles of the refractive index and the optical mode are plotted in Fig. 9.17. For lateral confinement, a 3-μm-wide ridge is etched into the top two layers. For this laser, calculated optical near and far fields are plotted in Fig. 4.23. The optical confinement factor is $\Gamma_o = 0.034$. The laser cavity is 450 μm long. The cleaved front facet has a power reflectivity of $R_f = 0.18$ and the back facet is coated with two periods of SiO_2/TiO_2 for high reflectivity ($R_b = 0.95$).

9.4.2   Optical Gain

The optical gain mechanism in InGaN quantum wells of real lasers is not yet fully understood. It may be strongly affected by a nonuniform Indium distribution. Internal polarization fields tend to separate quantum confined electrons and holes, thereby reducing optical gain and spontaneous emission. However, screening by electrons and holes is expected to suppress quantum well polarization fields at high power operation. The high carrier concentration also eliminates excitons. On the other hand, many-body models predict significant gain enhancement at high carrier concentrations [248]. Considering all the uncertainties in calculating the gain of real InGaN quantum wells, we here start with a simple free-carrier gain model, including a Lorentzian broadening function with 0.1 ps scattering time [34].

Table 9.4: Epitaxial Layer Structure and Room-Temperature Parameters of the Nitride Laser

Parameter Unit	d (nm)	N_{dop} (1/cm^3)	μ (cm^2/Vs)	n_r —	κ_L (W/cmK)
p-GaN (contact)	30	2×10^{20}	10	2.54	1.3
p-Al$_{0.14}$Ga$_{0.86}$N/GaN SL (cladding)	600	1×10^{20}	0.5	2.48	0.2
p-GaN (waveguide)	100	5×10^{18}	15	2.54	1.3
p-Al$_{0.2}$Ga$_{0.8}$N (stopper)	20	1×10^{19}	10	2.23	0.6
n-In$_{0.02}$Ga$_{0.98}$N (barrier)	10	7×10^{16}	850	2.61	0.7
n-In$_{0.15}$Ga$_{0.85}$N (quantum well)	4	7×10^{16}	300	3.0	0.2
n-In$_{0.02}$Ga$_{0.98}$N (barrier)	10	7×10^{16}	850	2.61	0.7
n-In$_{0.15}$Ga$_{0.85}$N (quantum well)	4	7×10^{16}	300	3.0	0.2
n-In$_{0.02}$Ga$_{0.98}$N (barrier)	10	7×10^{16}	850	2.61	0.7
n-GaN (waveguide)	100	7×10^{17}	550	2.54	1.3
n-Al$_{0.14}$Ga$_{0.86}$N/GaN SL (cladding)	1200	3×10^{18}	10	2.48	0.2
n-In$_{0.1}$Ga$_{0.9}$N (compliance)	100	3×10^{18}	390	2.98	1.0
n-GaN (substrate)	3000	3×10^{18}	410	2.54	1.3

Note. d, layer thickness; N_{dop}, doping density; μ, majority carrier mobility (low field); n_r, refractive index (wavelength 400 nm); κ_L, thermal conductivity; SL, super lattice.

Rectangular quantum wells are assumed. The resulting gain characteristics are plotted in Fig. 9.18 for different temperatures.

9.4.3 Comparison to Measurements

The common carrier mobility model of Eq. 3.28 results in good agreement between the simulated and the measured current–voltage (IV) characteristic (Fig. 9.19). This indicates that the contact resistance is of minor importance in this device. To find simultaneous agreement with the measured light–current (LI) characteristic, three fitted parameters, which are unknown for the present device, are employed: the internal modal loss α_i, the defect (SRH) recombination lifetime within the quantum wells τ_{qw}, and the thermal resistance R_{th}. The modal loss controls the slope efficiency, the lifetime affects the threshold current, and the thermal resistance has main impact on the power roll-off. The fitted values $\alpha_i = 12\,\mathrm{cm}^{-1}$, $\tau_{qw} = 0.5\,\mathrm{ns}$, and $R_{th} = 75\,\mathrm{K/W}$ result in an excellent reproduction of the experimental LI curve (Fig. 9.19). All three numbers are reasonable and confirm

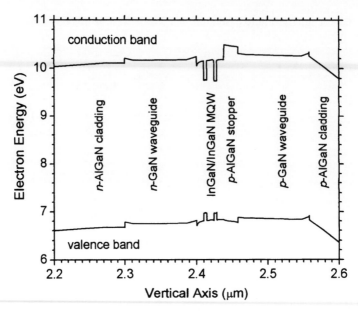

Figure 9.16: Energy band diagram of the nitride laser diode at high injection ($I = 540$ mA). The superlattice cladding layers are replaced by uniform AlGaN of average composition.

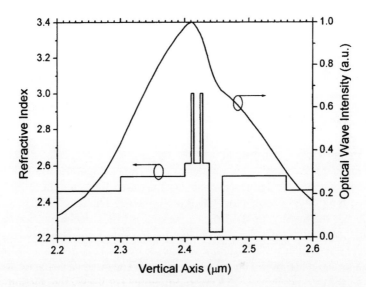

Figure 9.17: Vertical profile of refractive index and optical intensity.

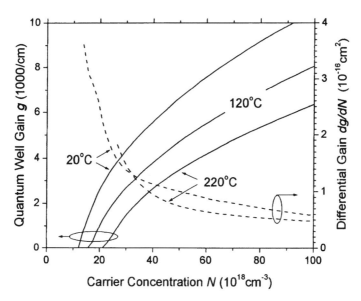

Figure 9.18: Gain (solid lines) and differential gain (dashed lines) as a function of carrier concentration for different temperatures (TE mode).

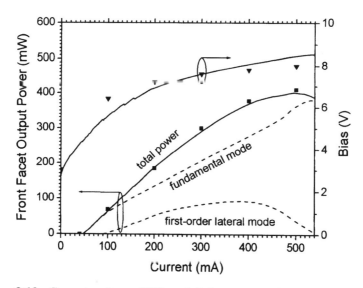

Figure 9.19: Current–voltage (IV) and light–current (LI) characteristics in continuous-wave operation: comparison of simulation (lines) and measurement (dots).

the accuracy of the laser model. In further agreement with the experimental observation [247], a first-order lateral laser mode emerges at a higher injection current in Fig. 9.19.

9.4.4 Power Limitations

The output power roll-off of laser diodes is typically attributed to the self-heating in continuous-wave operation. The data in Fig. 9.19 indicate that about $P_{heat} = 4$ W total heat power is generated at the output power maximum. Considering 75 K/W thermal resistance, the internal temperature rise can be estimated as $\Delta T = 300$ K (cf. Eq. (6.16)). The contour plot of the calculated temperature distribution $T(x, y)$ in Fig. 9.20 confirms this strong self-heating. The different heat sources are also plotted in this figure. Joule heat in the highly resistive p-doped regions dominates the power budget by far. The much smaller contribution from phonons generated by

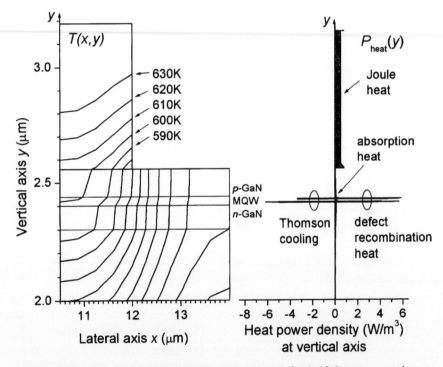

Figure 9.20: Left: temperature distribution $T(x, y)$ for half the cross-section at maximum output power (300 K ambient temperature). Right: Corresponding heat source profile $P_{heat}(y)$ at the laser axis ($x = 10.5\,\mu m$).

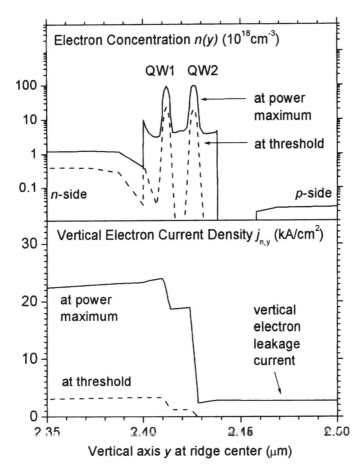

Figure 9.21: Vertical profile of electron concentration (top) and of the vertical electron current (bottom) at threshold (dashed line) and at maximum power (solid line).

defect recombination is partially compensated for by Thomson cooling, which represents the phonon absorption by carriers, for example, during thermionic emission from the quantum wells. Heat from photon absorption is of minor importance.

The strong self-heating reduces the optical gain substantially (Fig. 9.18), which is mainly caused by a wider Fermi spreading of carriers. Consequently, the carrier concentration increases with higher temperature (Fig. 9.21). The quantum well electron concentration is about 2×10^{19} cm^{-3} at threshold but it is substantially higher at the power maximum. This leads to enhanced recombination losses but, most of all, to an escalation of electron leakage from the quantum wells into the p-side. The bottom part of Fig. 9.21 plots the vertical electron current profile. At threshold, all electrons recombine within the quantum wells and no electron leaks into the p-side. At maximum power, a considerable portion of electrons escapes across the AlGaN stopper layer. This vertical electron leakage amounts to more than one-third of the total carrier losses at maximum power. As it prevents carriers from stimulated recombination, it is the main reason for the power roll-off [249].

9.4.5   Performance Optimization

Reduction of the self-heating seems to be the key to higher maximum output power. This can be achieved by lowering the heat power generation or by improved heat dissipation. The electrical resistance of the p-doped layers constitutes the main heat source but it cannot be easily reduced since higher p-doping causes a lower hole mobility. A wider ridge would substantially reduce the self-heating; however, the higher aspect ratio of the far field and the enhanced probability of higher order lasing modes are not desirable. Improved heat sinking has been demonstrated by replacing the sapphire substrate with copper [250]. In our simulation, we have used an external thermal resistance in addition to the simulated internal heat conduction. This external resistance represents the substrate and heat sink, which contribute 30 K/W to the total thermal resistance, $R_{th} = 75$ K/W. We now simply eliminate the external thermal resistance, which reduces the total thermal resistance to 45 K/W. The resulting LI characteristic exhibits almost double the maximum output power (Fig. 9.22). The quantum well temperature at maximum power is about the same as that with the original laser.

Suppression of electron leakage may be achieved by using an AlGaN stopper layer with a larger band gap. Such a situation is simulated by assuming an AlGaN electron stopper layer with 25% aluminum. The larger conduction band offset results in a very slight increase of the maximum output power (Fig. 9.22). This is due to reduced electron leakage since all other material parameters of the stopper layer remain unchanged. However, thermal and electrical conductivities decrease with higher Al content, which is expected to result in lower maximum power. Antireflection coating of the front facet ($R_f = 0.02$) has a stronger effect on the maximum power in Fig. 9.22.

Further Reading

- S. Nakamura and S. F. Chichibu (Eds.), *Introduction to Nitride Semiconductor Blue Lasers and Light-Emitting Diodes*, Taylor & Francis, London 2000.

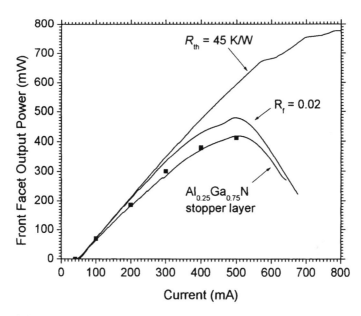

Figure 9.22: Light–current characteristics resulting from laser improvements (dotted, original measurement; R_{th}, thermal resistance; R_f, front facet reflectance).

- S. J. Pearton (Ed.), *GaN and Related Materials II*, Gordon and Breach, Amsterdam, 2000.

- E. F. Schubert, *Light-Emitting Diodes*, Cambridge Univ. Press, Cambridge, UK, 2003.

Chapter 10

Electroabsorption Modulator

A multiquantum well (MQW) electroabsorption modulator that employs the quantum confined Stark effect to absorb light near 1.55-μm wavelength is analyzed. Strained layers are utilized in this device to obtain polarization insensitive absorption. This is accompanied by strong valence band mixing near the Γ point, rendering common effective mass approximations invalid. The optical transmission is investigated as a function of wavelength and bias.

10.1 Introduction

In optical communication systems, signals are transmitted by modulating the phase or the intensity of an optical wave. The modulation can be performed directly by the laser diode that generates the optical wave, e.g., by varying the injection current. Such direct modulation is accompanied by variations of the carrier concentration in the active region, which leads to fluctuations of the refractive index and wavelength (chirp). To reduce the chirp, modulation is often performed externally, after the optical wave leaves the laser diode. There are generally two types of external modulators, electro-optic and electroabsorption modulators. The first type is based on the linear electro-optic effect; i.e., on the change of the refractive index by an electric field. By varying the electric field, the optical phase is modulated. This phase modulation can be transformed into intensity modulation using Mach–Zehnder interferometers. Dielectric materials such as $LiNbO_3$ are most widely used in electro-optic modulators. They require a relatively high reverse bias of more than 3 V to achieve sufficiently strong modulation. The ratio between highest and lowest signal intensity (extinction ratio) is typically in the range from 15 to 20 dB.

Electroabsorption modulators can achieve the same extinction ratio at lower voltage. They are based on the electroabsorption effect, i.e., on the change of the absorption coefficient by an electric field. In bulk semiconductors, the absorption edge moves to lower energies with increasing electric field due to a combination of band-to-band absorption and tunneling processes (Franz–Keldysh effect). In quantum wells, the transition energy between confined energy levels for electrons and holes is reduced as an electric field is applied in the growth direction (quantum confined Stark effect, QCSE). This effect is illustrated in Fig. 10.1. As

213

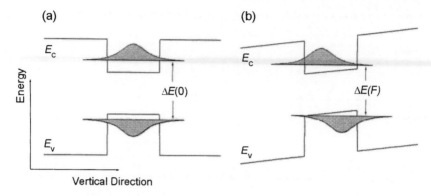

Figure 10.1: Illustration of the quantum confined Stark effect: quantum well band edges, quantum levels, and wave functions (a) without and (b) with applied electric field F.

the field is increased, the overlap of electron and hole wave functions is reduced, thereby decreasing the absorption strength at the transition energy. Thick quantum wells are advantageous for high field sensitivity (high modulation efficiency), whereas thin quantum wells give stronger absorption. The formation of bound electron–hole pairs (excitons) in the quantum well further enhances the absorption [157]. Compared to bulk materials, quantum well QCSE-type modulators have higher modulation efficiencies (lower drive voltage); however, Franz–Keldysh type modulators are less wavelength sensitive (larger optical bandwidth).

In this chapter, we present the simulation and analysis of a multiquantum well QCSE-type modulator [251]. The InGaAsP/InP device is designed for fiberoptic applications near 1550-nm wavelength. The ridge–waveguide device was grown on semi-insulating (SI) InP; the layer structure is given in Table 10.1 [252]. Section 10.2 evaluates the quantum well design, Section 10.3 investigates the optical waveguide, and Section 10.4 analyzes optical transmission characteristics.

10.2 Multiquantum Well Active Region

Polarization independent performance is an important requirement in many applications of electroabsorption modulators. In the InGaAsP/InP material system, TE polarized waves mainly generate heavy holes, whereas TM polarized waves only generate light holes. The corresponding transition matrix elements result in the absorption coefficients

$$\alpha_{\text{TE}} = \frac{1}{4}\alpha_{\text{lh}} + \frac{3}{4}\alpha_{\text{hh}} \tag{10.1}$$

$$\alpha_{\text{TM}} = \alpha_{\text{lh}}, \tag{10.2}$$

Table 10.1: Layer Structure and Parameters of the Electroabsorption Modulator with a Total of 10 Quantum Wells and 11 Barriers

Parameter Unit	d (nm)	N_{dop} $(1/cm^3)$	n_r —
p-InP	500	1×10^{18}	3.164
p-InP	1000	5×10^{17}	3.164
i-InP	150	5×10^{15}	3.164
i-In$_{0.923}$Ga$_{0.077}$As$_{0.325}$P$_{0.675}$(barrier)	7.6	5×10^{15}	3.305
i-In$_{0.485}$Ga$_{0.515}$As$_{0.979}$P$_{0.021}$(well)	10.4	5×10^{15}	3.576
i-In$_{0.923}$Ga$_{0.077}$As$_{0.325}$P$_{0.675}$(barrier)	7.6	5×10^{15}	3.305
. . .			
i-In$_{0.485}$Ga$_{0.515}$As$_{0.979}$P$_{0.021}$(well)	10.4	5×10^{15}	3.576
i-In$_{0.923}$Ga$_{0.077}$As$_{0.325}$P$_{0.675}$(barrier)	7.6	5×10^{15}	3.305
i-InP	30	5×10^{15}	3.164
n-InP (cladding)	300	3×10^{18}	3.164
n-In$_{0.72}$Ga$_{0.28}$As$_{0.605}$P$_{0.395}$ (etch-stop)	20	3×10^{18}	3.398
n-InP (conducting)	500	3×10^{18}	3.164
InP (SI substrate)	65 μm	—	3.164

Note. d, layer thickness; N_{dop}, doping density; n_r, refractive index at 1550-nm wavelength. Intrinsic (i) layers are assumed to exhibit low n-type background doping.

where α_{hh} and α_{lh} are the band-to-band absorption coefficients for the heavy- and light-hole bands, respectively. Thus, $\alpha_{hh} = \alpha_{lh}$ is a basic requirement for polarization insensitive modulation. In unstrained quantum wells, the transition energy for heavy holes is lower than for light holes due to the different effective mass (cf. Fig. 2.15). To align the absorption edges, the light-hole band gap needs to be reduced by using tensile strain in the quantum well. The barrier layers are compressively strained for strain compensation. The active region of our device comprises ten 10.4-nm-thick InGaAsP quantum wells that exhibit 0.37% tensile strain (Table 10.1). The InGaAsP barriers are 7.6 nm thick with 0.5% compressive strain. Calculated absorption spectra are plotted in Fig. 10.2. Higher reverse voltage reduces the quantum well transition energy, and it shifts the absorption edge to longer wavelengths. For both voltages, the absorption edges are almost identical for TE and TM polarized waves. Correct calculation of the absorption edge shape is essential to find good agreement with modulation measurements. This shape depends on the energy broadening model used (cf. Section 5.1.2). We here use a Lorentzian function with 0.2-ps

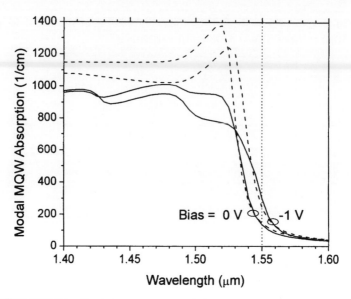

Figure 10.2: MQW absorption spectra for TE (solid line) and TM (dashed line) polarization, respectively, at two different voltages (fundamental mode).

intraband scattering time, which results in good agreement with transmission measurements [253].

The degeneration of heavy- and light-hole levels in our quantum wells is accompanied by an enhanced interaction of both particles. This interaction causes a substantial deformation of the valence subbands (valence band mixing). Figure 10.3 shows the resulting nonparabolic shape of the valence bands; i.e., the effective mass model of parabolic bands cannot be applied here. The fundamental absorption edges in Fig. 10.2 are dominated by the first subbands (HH1, LH1); higher subbands cause additional steps in the absorption spectrum.

Band-to-band absorption can be understood as negative gain (cf. Chap. 5). Both the gain g and the refractive index n_r change with the average quantum well carrier concentration N. The ratio of these changes gives the chirp factor (linewidth enhancement factor)

$$\alpha_{\text{chirp}} = -\frac{4\pi}{\lambda_0} \frac{dn_r/dN}{dg/dN}. \tag{10.3}$$

A slightly negative chirp factor is ideal for digital fiber optic communication systems [254]. It is typically between 4 and 6 for semiconductor lasers, while it is close to 0 or even negative for electroabsorption modulators. Figure 10.4 compares both cases for our quantum well. High carrier concentrations are typical

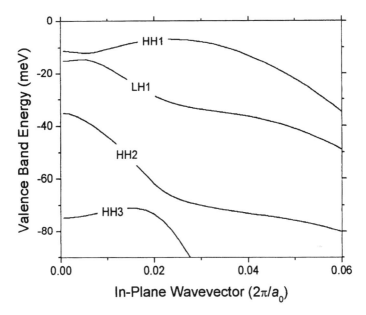

Figure 10.3: Valence subbands for polarization insensitive design (HH#, heavy-hole subbands; LH#, light-hole subbands; $F = 0$).

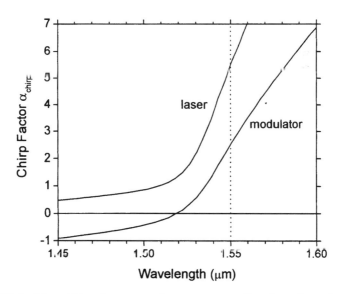

Figure 10.4: Linewidth enhancement factor α_{chirp} as a function of wavelength for low (modulator) and high carrier concentration (laser), respectively, at zero field.

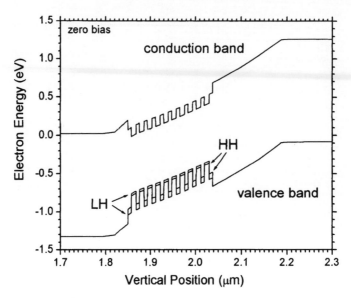

Figure 10.5: Energy band diagram of the MQW active region with edges of the heavy-hole band (HH), light-hole band (LH), and conduction band.

for laser diodes, and they do not allow for negative chirp factors. Modulators are reverse biased and exhibit a much lower carrier concentration in the quantum wells. In Fig. 10.4, the transition wavelength to negative chirp factors is about 1.52 μm, which is near the absorption peak (cf. Fig. 10.2). Since the modulator rather operates in the tail region of the absorption spectrum, our chirp factor is expected to be slightly positive.

The energy band diagram of the multiquantum well active region is plotted in Fig. 10.5 at zero bias, indicating a significant built-in electrostatic field. Including the undoped InP regions, the total intrinsic region thickness of the *pin* structure is about 370 nm. Figures 10.6 and 10.7 show the internal field and the conduction band edge, respectively, for different applied voltages. At 1 V forward bias, the MQW field vanishes and the band edges are flat. At reverse bias, the internal field approaches 10^5 V/cm.

10.3 Optical Waveguide

Our device utilizes a ridge structure for waveguiding with a 2-μm-wide ridge etched beyond the intrinsic region. The fundamental optical mode is shown in Fig. 10.8 for half of the device cross section. The narrow ridge gives a more circular mode profile for better coupling to optical fibers. The vertical refractive

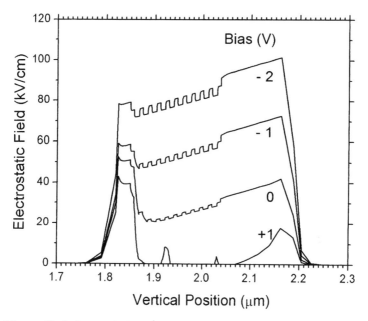

Figure 10.6: Internal electrostatic field at different applied voltages.

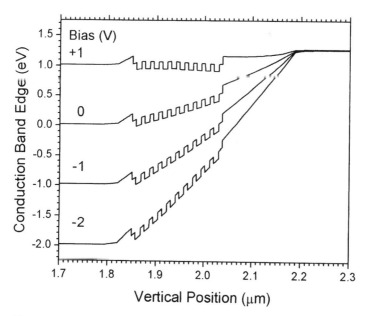

Figure 10.7: Conduction band edge at different applied voltages.

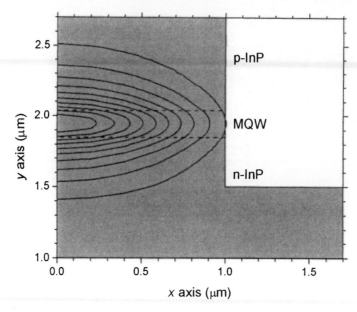

Figure 10.8: Half cross section of the modulator waveguide with fundamental optical mode (the left border is the vertical symmetry plane of the device).

index profile is plotted in Fig. 10.9 together with the mode intensity. The optical confinement factor is $\Gamma_o = 0.2$. Stronger vertical confinement is desired for low voltage operation; however, it causes higher fiber-coupling losses. The optical model includes absorption losses by free carriers that are roughly proportional to the carrier concentration and only relevant in p-doped regions ($\alpha_p = k_p p$ with $k_p = 25 \times 10^{-18} \, \text{cm}^2$). Figure 10.9 shows the resulting free-carrier absorption profile. Due to the small overlap with the optical mode, the corresponding modal loss is only 1.5 cm^{-1}. Additional optical losses are caused by photon scattering at the rough sidewalls of the etched waveguide; however, scattering losses are hard to calculate. The strongest contribution to the modal optical loss in our device results from the residual band-to-band absorption within the quantum wells (Fig. 10.2).

10.4 Transmission Analysis

The optical transmission of the modulator is a function of the wavelength λ_0 and of the applied voltage V

$$T(\lambda_0, V) = \gamma_c^2 (1 - R)^2 \exp\left[-\Gamma_o \alpha_{qw}(\lambda_0, V) L\right] \qquad (10.4)$$

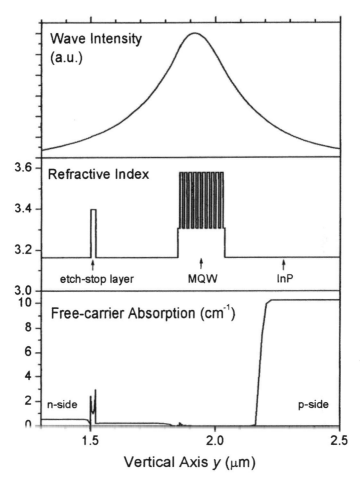

Figure 10.9: Vertical profile of optical mode, refractive index, and free-carrier absorption.

with the device length $L = 300\,\mu m$. The fiber-coupling coefficient γ_c gives the overlap of the optical modes of fiber and modulator. The facet reflectance R is kept low by antireflection coating. In our case, the optical loss at each facet is $\gamma_c(1 - R) \approx 0.5\,(-3\,dB)$. The modal absorption $\Gamma_o\alpha_{qw}$ is shown in Fig. 10.2. The corresponding transmission characteristics are given in Fig. 10.10 as a function of wavelength with the bias as parameter. With increasing reverse bias, the absorption edge moves toward longer wavelength, and the maximum absorption is reduced, as seen in Fig. 10.2. Three different wavelengths that shall be compared in the following are marked in Fig. 10.10. At 1560 nm, measured results show good

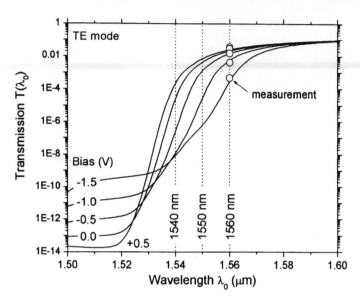

Figure 10.10: Optical transmission vs wavelength with the bias as parameter. The circles give measured results at the same bias values [252].

agreement with the simulation for all bias values. At 1550-nm wavelength and zero bias, the residual transmission is $4.5 \times 10^{-3}(-23\,\text{dB})$. A bias of -1 V gives 3×10^{-5} transmission ($-45\,\text{dB}$), corresponding to an extinction ratio of 0.0066 ($-22\,\text{dB}$). These numbers change with the wavelength. The wavelength sensitivity of electroabsorption modulators depends on the distance to the absorption edge. In our case, it is small for wavelengths below 1530 nm and above 1570 nm. For three wavelengths near the absorption edge, Fig. 10.11 displays the transmission as a function of voltage. As discussed above, the transmission is reduced with higher reverse voltage. Longer wavelengths experience less absorption at any given bias. At 1540 nm, the highest possible transmission is below 10^{-3}, and saturation occurs at stronger reverse bias as the absorption plateau is reached (cf. Fig. 10.10). The comparison between the TE and TM modes in Fig. 10.11 illustrates that the polarization sensitivity depends on wavelength and bias.

The first derivative dT/dV is given in Fig. 10.12. At 1560 nm, the steepest slope of $T(V)$ occurs at about -1 V reverse voltage, which gives the highest efficiency of analog modulation. Shorter wavelengths experience less efficient modulation. For 1550 nm, the optimum bias is close to 0 V. The low modulation efficiency and the low total transmission disqualify 1540 nm as a usable wavelength for this device.

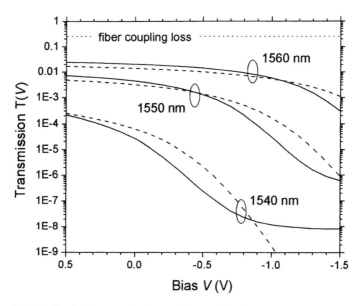

Figure 10.11: Optical transmission vs bias with the wavelength as parameter (solid lines, TE mode; dashed lines, TM mode).

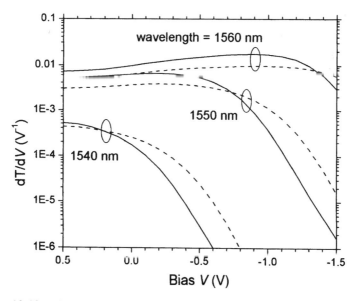

Figure 10.12: First derivative of the optical transmission vs bias with the wavelength as parameter (solid lines, TE mode; dashed lines, TM mode).

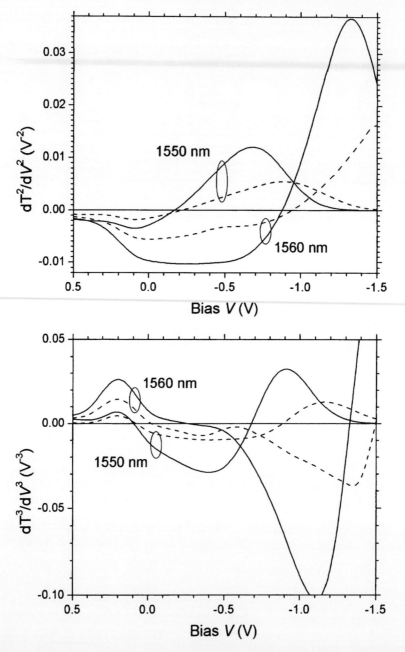

Figure 10.13: Second (top) and third (bottom) derivatives of the transmission $T(V)$ as a function of bias for two different wavelengths (solid lines, TE mode; dashed lines, TM mode).

Linearity is another main requirement in analog applications of electroabsorption modulators. Both the second derivative d^2T/dV^2 and the third derivative d^3T/dV^3 of the transmission function $T(V)$ should be as small as possible. However, this is hard to achieve at the same bias as shown in Fig. 10.13. At 1550-nm wavelength, dT^2/dV^2 is zero near -0.2 V; however, the third derivative is close to a maximum at this bias. A zero third derivative at 0.1 V forward bias is accompanied by a relatively small second derivative. Thus, for 1550-nm signal wavelength, the optimum region for analog operation with our device seems to be near zero bias. Signals at 1560 nm experience slightly higher transmission and modulation efficiency but also stronger nonlinearity.

Further Reading

- M. Fukuda, *Optical Semiconductor Devices*, Wiley, New York, 1999.

- A. Ahland, D. Schulz, and E. Voges, Efficient modeling of the optical properties of MQW modulators on InGaAsP with absorption edge merging, *IEEE J. Quant. Electron.*, **34**, 1597–1603 (1998).

- G. L. Li, C. K. Sun, S. A. Pappert, W. X. Chen, and P. K. L. Yu, Ultrahigh-speed traveling-wave electroabsorption modulator — Design and analysis, *IEEE Trans. Microwave Theory Techniques*, **47**, 1177–1183 (1999).

- T. Aizawa, K. G. Ravikumar, S. Suzaki, T. Watanabe, and R. Yamauchi, Polarization-independent quantum-confined Stark effect in an InGaAs/InP tensile-strained quantum well, *IEEE J. Quant. Electron.*, **30**, 585–592 (1994).

- D. A. B. Miller, D. S. Chemla, T. C. Damen, A. C. Gossard, W. Wiegmann, T. H. Wood, and C. A. Burrus, Electric field dependence of optical absorption near the band gap of quantum well structures, *Phys. Rev. B*, **32**, 1043–1060 (1985).

Chapter 11

Amplification
Photodetector

The concept of waveguide photodetection with integrated amplification is
evaluated by two-dimensional device simulation. A bulk GaAs photodetector
region is combined with GaAs quantum wells for amplification. The current
flow in the three-terminal device is analyzed. The net optical gain is calculated
for different waveguide modes, identifying the preferred mode of operation.
For this mode, the detector responsivity is shown to scale with the device
length, reaching a quantum efficiency larger than 100%.

11.1 Introduction

The purpose of any photodetector is to convert light (photons) into electric current
(electrons). Photodetectors are among the most common optoelectronic devices;
they automatically open supermarket doors, detect the signal from infrared remote
controls, and record pictures in modern cameras. Photodetectors are fabricated
from various semiconductor materials, since the band gap needs to be smaller than
the energy of the photons detected. Photon absorption generates electron–hole
pairs which are subsequently separated by the applied electrical field. Depending
on the desired use, photodetectors are designed in many different ways. We investigate here a waveguide photodetector that looks similar to an in-plane laser, only
the light is *injected* through the antireflection coated front facet of the waveguide.
The electrical field is applied in the vertical direction so that photon flux and
carrier flux are perpendicular to each other. Such devices are suitable for high
data-rate applications since thin absorption layers allow for short transit times of
the generated carriers. The detection efficiency can be increased by using long
waveguides. Ideally, every photon should generate one electron–hole pair, giving
100% quantum efficiency. However, only a small fraction of the optical signal
is usually converted into an electrical signal. To obtain stronger electrical output power, the optical input power may be increased by preamplification. This
method is limited by saturation effects, as intense light leads to a large carrier
concentration in the absorption layer, which changes the behavior of the detector.
A solution to this problem was recently proposed by integrating the detector and
amplifier [255].

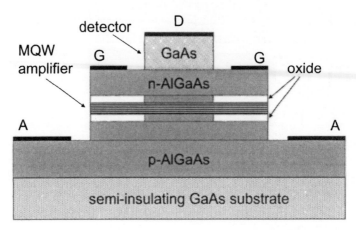

Figure 11.1: Cross section of the amplification photodetector (D, detector contact; G, ground contact; A, amplifier contact).

The amplification photodetector simultaneously amplifies and absorbs the incoming light. The schematic cross section of such a device is shown in Fig. 11.1. The top GaAs ridge is the detecting layer, all other semiconductor layers exhibit a larger band gap. Light amplification is provided by multiple GaAs quantum wells. The amplifier *pn*-diode is operated at forward bias, whereas the detector diode is reverse biased. Electrical and optical confinement in the center of the device is provided by the ridge–waveguide structure as well as by lateral oxidation. The continuous amplification of the optical wave makes it possible to maintain a constant optical power along the waveguide despite the absorption. With long devices, much higher output power than in conventional photodetectors can be achieved. Details of the device design are evaluated in the following using two-dimensional APSYS simulations.

11.2 Device Structure and Material Properties

The layer structure of the device is listed in Table 11.1. Refractive index data for AlGaAs are obtained from the Afromowitz model described in Section 4.2.2. Oxidation layers exhibit a high Al mole fraction, which lowers the refractive index. The vertical index profile is plotted in Fig. 11.2 together with several vertical waveguide modes. The behavior of these optical modes will be analyzed in the next section. Lateral mode confinement is provided by the top ridge as well as by the oxide layers. A value of 1.7 is used for the refractive index of aluminum oxide. The width of the oxide aperture is set equal to the ridge width of $2 \, \mu m$.

The $Al_x Ga_{1-x} As$ system converts from a direct semiconductor for $x < 0.42$ to an indirect semiconductor for $x > 0.42$ (see Fig. 2.16). Our device contains

Table 11.1: Epitaxial Layer Structure and Parameters of the Amplification Photodetector

Parameter Unit	d (nm)	N_{dop} $(1/cm^3)$	n_r —
i-GaAs (absorption)	300	1×10^{16}	3.65
n-$Al_{0.15}Ga_{0.85}As$ (cladding)	100	5×10^{18}	3.52
n-$Al_{0.2}Ga_{0.8}As$ (cladding)	300	1×10^{18}	3.48
n-$Al_xGa_{1-x}As$ (grading $x = 0.9$ to 0.2)	16.5	1×10^{18}	3.05 to 3.48
n-$Al_{0.9}Ga_{0.1}As$ (oxidation)	6	1×10^{18}	3.05
n-$Al_{0.98}Ga_{0.02}As$ (oxidation)	32	1×10^{18}	3.00
n-$Al_{0.9}Ga_{0.1}As$ (oxidation)	6	1×10^{18}	3.05
n-$Al_xGa_{1-x}As$ (grading $x = 0.15$ to 0.9)	18.2	1×10^{18}	3.52 to 3.05
n-$Al_{0.15}Ga_{0.85}As$ (waveguide)	90	1×10^{18}	3.52
i-$Al_{0.15}Ga_{0.85}As$ (waveguide)	10	1×10^{16}	3.52
i-GaAs (quantum well)	8	1×10^{16}	3.65
i-$Al_{0.15}Ga_{0.85}As$ (barrier)	8	1×10^{16}	3.52
i-GaAs (quantum well)	8	1×10^{16}	3.65
i-$Al_{0.15}Ga_{0.85}As$ (barrier)	8	1×10^{16}	3.52
i-GaAs (quantum well)	8	1×10^{16}	3.65
i-$Al_{0.15}Ga_{0.85}As$ (barrier)	8	1×10^{16}	3.52
i-GaAs (quantum well)	8	1×10^{16}	3.65
i-$Al_{0.15}Ga_{0.85}As$ (waveguide)	68	1×10^{16}	3.52
p-$Al_{0.2}Ga_{0.8}As$ (waveguide)	20	5×10^{17}	3.48
p-$Al_xGa_{1-x}As$ (grading $x = 0.9$ to 0.2)	18.2	1×10^{18}	3.05 to 3.48
p-$Al_{0.9}Ga_{0.1}As$ (oxidation)	6	3×10^{17}	3.05
p-$Al_{0.98}Ga_{0.02}As$ (oxidation)	32	3×10^{17}	3.00
p-$Al_{0.9}Ga_{0.1}As$ (oxidation)	6	3×10^{17}	3.05
p-$Al_xGa_{1-x}As$ (grading $x = 0.2$ to 0.9)	16.5	2×10^{18}	3.48 to 3.05
p-$Al_{0.2}Ga_{0.8}As$ (cladding)	300	5×10^{17}	3.48
p-$Al_{0.2}Ga_{0.8}As$ (cladding)	200	2×10^{18}	3.48
p-$Al_{0.2}Ga_{0.8}As$ (cladding)	100	5×10^{18}	3.48
p-$Al_{0.5}Ga_{0.5}As$ (sublayer)	3000	1×10^{16}	3.29

Note. d, layer thickness; N_{dop}, doping density; n_r, refractive index at 840-nm wavelength. Intrinsic (i) layers are assumed to exhibit low p-type background doping.

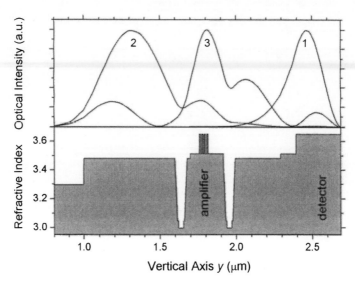

Figure 11.2: Vertical profile for different waveguide modes (top) and for the refractive index at 840-nm wavelength (bottom).

both types of materials. The hole transport is hardly affected by this transition. However, most electrons entering the n-doped oxidation layers are transferred from the Γ valley into the X side valley of the conduction band. Due to the high Al mole fraction, a negligible number of electrons are still traveling in the Γ valley. Thus, the lower band edge is used in calculating the carrier transport across heterointerfaces. The common band offset ratio of $\Delta E_c / \Delta E_g = 0.65$ is employed at all interfaces. Figure 11.3 shows the band diagram at the vertical axis of the device. Due to the grading and doping profile used, the valence band edge of the oxidation layers is almost flat on the p-doped side, and it hardly affects the hole injection into the multiquantum well (MQW) amplification region. The GaAs detector region is reverse biased.

The conduction bands of the active GaAs layers are assumed parabolic and the nonparabolic valence bands are calculated by the two-band $\vec{k} \cdot \vec{p}$ method (cf. Section 2.2). Quantum well gain spectra are plotted in Fig. 11.4 for different carrier concentrations (cf. Section 5.1.2). At 840-nm wavelength, amplifier gain on the order of $g_{\mathrm{amp}} = 2000\,\mathrm{cm}^{-1}$ is anticipated for the TE mode. For low carrier concentration (reverse bias), the calculated absorption spectrum of the bulk GaAs layer is also shown in Fig. 11.4, and it is in good agreement with the experimental data in Fig. 4.4. At 840-nm wavelength, the absorption coefficient is $\alpha_{\mathrm{det}} \approx 6000\,\mathrm{cm}^{-1}$. The total amplification layer thickness of 32 nm is much smaller than

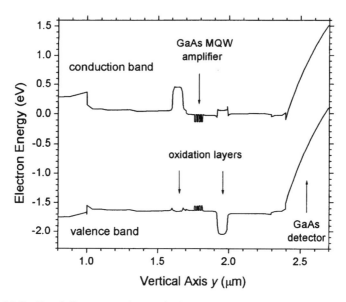

Figure 11.3: Band diagram at the vertical axis with $V_D = -1$ V detector voltage and $I_A = 20$ mA amplifier current.

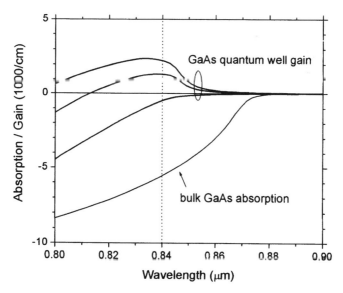

Figure 11.4: Gain and absorption spectra for quantum well and ridge, respectively. Gain spectra are calculated at three carrier concentrations: 2, 4, and 6×10^{18} cm^{-3}; the bulk absorption is calculated at 10^{16} cm^{-3}.

the absorption layer thickness of 300 nm. Thus, optical waveguide modes that exhibit little overlap with the detector region are preferred.

11.3 Waveguide Mode Analysis

Three main vertical modes of our device are shown in Fig. 11.2, and mode 2 is given as a 2D contour plot in Fig. 11.5. Mode 1 is mainly located in the detector region, mode 3 peaks in the amplifier region, and mode 2 does not show much overlap with any active layer. The optical confinement factors of these three modes are given in Table 11.2 for both active regions. The total modal gain of each mode can be estimated as

$$g_{mode} = \Gamma_{amp}g_{amp} - \Gamma_{det}\alpha_{det}. \tag{11.1}$$

The normalized optical power in the travel direction is given in Fig. 11.6 for all three modes. Mode 1 is completely absorbed after a short travel distance. Mode 3 also suffers from net absorption. Only mode 2 exhibits positive gain. Therefore, we focus on mode 2 in the following.

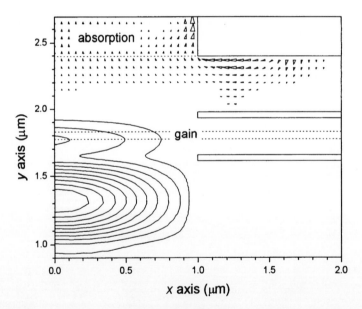

Figure 11.5: Contours: 2D intensity distribution of mode 2. Arrows: current density at $I_A = 20\,mA$ amplifier current, $V_D = -1\,V$ detector voltage, and $P_2 = 100\,mW$ modal power (the arrows scale with the current density and disappear for less than 10% of the maximum).

Table 11.2: Optical Confinement Factors Γ_{amp} and Γ_{det} of the Vertical Modes in Fig. 11.2 for the Amplification and Detection Layers, Respectively

Mode	Γ_{amp}	Γ_{det}
1	0.00014	0.6769
2	0.0165	0.0028
3	0.0684	0.0524

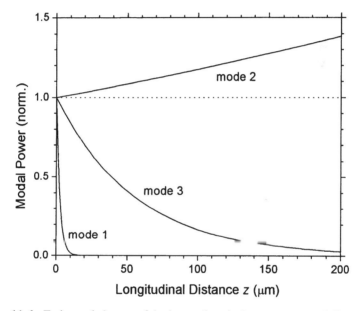

Figure 11.6: Estimated change of the internal optical power vs travel distance for all three modes from Fig. 11.2.

The net modal gain spectrum is plotted in Fig. 11.7 as a function of the amplifier current. At the optimum wavelength of 840 nm, mode 2 shows zero net gain for 21-mA amplifier current, which gives constant optical power along the waveguide. The other two modes do not reach zero net gain for any reasonable pumping current. Positive gain is needed to overcome all other optical losses, e.g., photon scattering losses, which are assumed to be 5 cm^{-1}.

Many different waveguide modes may be excited by coupling light into the device, e.g., from an optical fiber. Most of these modes are quickly absorbed by

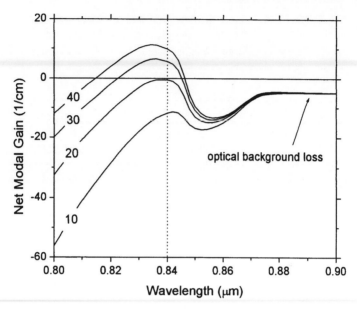

Figure 11.7: Modal gain spectrum for mode 2 from Fig. 11.2 at different amplifier currents (mA).

the thick detector region. Only those modes are useful in our device for which about the same number of photons is generated in the quantum wells and absorbed in the detector. This transfer of photons is the basic idea of the amplification photodetector. Photons multiplied by stimulated emission stay within the same mode. Thus, in our device example, the incoming signal needs to be mainly coupled into modes guided by the lower cladding layers, similar to mode 2.

11.4 Detector Responsivity

In our 2D simulation, we now investigate the detector's response to waveguide mode 2 for the special case of constant longitudinal mode power ($I_A = 21$ mA). Figure 11.8 shows a vector plot of the current density distribution at low optical power ($P_2 = 10$ mW, $V_D = -1$ V). The plot reveals current crowding at the edges of the oxide aperture as well as lateral current leakage within the MQW amplifier region. The detector generates about $I_D \approx 5$ mA current, which is still smaller than I_A so that the amplifier dominates the direction of the current flow at the ground contact. This situation changes with higher modal power. At $P_2 = 100$ mW, the detector current is close to 50 mA, and the current flow at the ground electrode reverses (cf. Fig. 11.5).

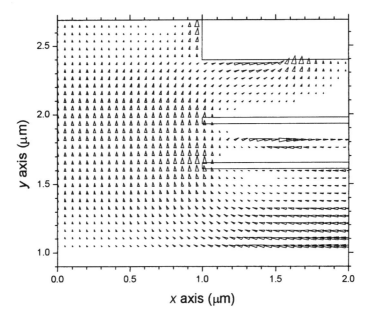

Figure 11.8: Vector plot of the current density at $I_A = 21$ mA amplifier current and $P_2 = 10$ mW modal power (the size of the arrows scales with the current density).

Figure 11.9 plots the photocurrent as a function of the modal power for two different detector lengths L. Due to its little overlap with the absorption layer, mode 2 shows a very linear slope without any detector saturation effect up to 100-mW input power.[1] The detector responsivity is $dI_D/dP = 0.46$ A/W at L $= 300$ μm, and it is twice as high with double the device length. The quantum efficiency is given by

$$\eta_{\text{det}} = \frac{dI_D}{dP} \frac{\hbar\omega}{q}, \tag{11.2}$$

and it surpasses 100% for L > 440 μm in our case ($\hbar\omega/q = 1.476$ W/A at 840-nm wavelength). 200% efficiency would be reached with L $- 880$ μm, i.e., each incident photon then generates on average two electron-hole pairs in the detector layer, due to photon multiplication in the amplifier region.

---

[1] Saturation effects depend on the longitudinal position and require 3D simulations.

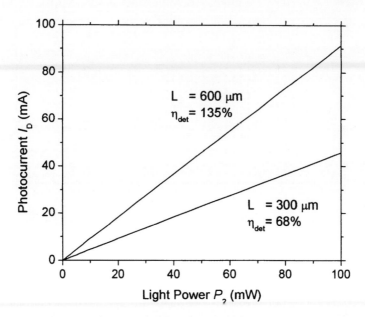

Figure 11.9: Photocurrent as a function of second mode power at 840-nm wavelength with $I_A = 21$ mA pump current and $V_D = -1$ V detector voltage. The quantum efficiency η_{det} scales with the detector length L since the modal power is constant in the longitudinal direction.

Further Reading

- S. Donati, *Photodetectors — Devices, Circuits, and Applications*, Prentice-Hall, Upper Saddle River, NJ, 2000.

- H. Ghafouri-Shiraz, *Semiconductor Optical Amplifiers*, Wiley, New York, 1995.

- P. Bhattacharya, *Semiconductor Optoelectronic Devices*, 2nd ed., Prentice-Hall, Upper Saddle River, NJ, 1997.

Appendix A

Constants and Units

A.1 Physical Constants

Speed of light in free space	c_0	299.8×10^6 m/s
Planck constant	h	6.6261×10^{-34} Js
		$= 4.1357 \times 10^{-15}$ eV s
Reduced Planck constant	$\hbar$	1.0546×10^{-34} Js
		$= 6.5822 \times 10^{-16}$ eV s
Boltzmann constant	k_B	1.3807×10^{-23} J/K
Thermal energy at $T = 300$ K	$k_B T$	25.853 meV
Free electron mass	m_0	9.1094×10^{-31} kg
Elementary charge	q	1.6022×10^{-19} C
Permeability in free space	μ_0	$4\pi \times 10^7$ N/A^2
		$= 12.566 \times 10^7$ N/A^2
Permittivity in free space	ϵ_0	$1/\mu_0 c_0^2 = 8.854 \times 10^{-12}$ F/m

A.2 Unit Conversion

Length, distance	1 m $= 39.37$ in $= 10^{10}$ Å (Angstrom)
Temperature	1 K (Kelvin) $= 1°C - 273.15$
Frequency	1 Hz (Hertz) $= $ s^{-1}
Charge	1 C (Coulomb) $= 1$ A s
Resistance	1 Ω (Ohm) $= 1$ V A^{-1}
Capacity	1 F (Farad) $= 1$ A s V^{-1}
Force	1 N (Newton) $= 1$ kg m s^{-2} $= 10^5$ dyn
Power	1 W (Watt) $= 1$ V A
Energy	1 J (Joule) $= 1$ W s $= 1$ N m $= 10^7$ erg
	$= 6.242 \times 10^{18}$ eV
Pressure	1 Pa (Pascal) $= 1$ N m^{-2} $= 10^{-5}$ bar
	$= 9.872 \times 10^{-6}$ atm
Magnetic induction	1 T (Tesla) $=$ Wb (Weber) m^{-2} $=$ V s m^{-2}
	$= 10^4$ G (Gauss)

Appendix B

Basic Mathematical Relations

B.1 Coordinate Systems

The most common coordinates are the Cartesian coordinates (x, y, z) which relate to the three perpendicular directions in space (Fig. B.1). The origin of the coordinate system $(0, 0, 0)$ is often chosen to be located at an interface or at some other convenient position. Alternatively, the location of a point in space can be given by it's distance R to the origin and by the angles Θ and Φ as shown in Fig. B.1. The set (R, Θ, Φ) is called spherical coordinate system. It is related to the Cartesian coordinates by

$$x = R \sin \Theta \cos \Phi \qquad (B.1)$$
$$y = R \sin \Theta \sin \Phi \qquad (B.2)$$
$$z = R \cos \Theta. \qquad (B.3)$$

and it is used, for instance, to describe the laser far field (Section 4.13). The third relevant system is the cylindrical coordinate system (r, Φ, z) with

$$x = r \cos \Phi \qquad (B.4)$$
$$y = r \sin \Phi. \qquad (B.5)$$

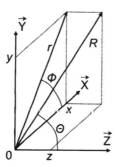

Figure B.1: Illustration of coordinate systems.

The cylindrical system is employed to describe cylindrical structures such as optical fibers and vertical-cavity surface-emitting lasers (cf. Chap. 8).

B.2 Vector and Matrix Analysis

A vector is a physical quantity like the velocity $\vec{v}$ having a magnitude and a direction. The vector is usually given by the three vector components (v_x, v_y, v_z) in the axis direction of the Cartesian coordinate system (x, y, z). As shown in Fig. B.1, these three directions can also be described by the unit vectors $\vec{X}, \vec{Y}, \vec{Z}$, respectively,

$$\vec{v} = v_x\vec{X} + v_y\vec{Y} + v_z\vec{Z}. \tag{B.6}$$

The magnitude of the vector $\vec{v}$ is given by

$$|\vec{v}| = v = \sqrt{v_x^2 + v_y^2 + v_z^2}. \tag{B.7}$$

The scalar product of two vectors $\vec{v}$ and $\vec{u}$ is calculated by

$$\vec{v} \cdot \vec{u} = v_x u_x + v_y u_y + v_z u_z, \tag{B.8}$$

and their vector product by

$$\vec{v} \times \vec{u} = (v_y u_z - v_z u_y)\vec{X} + (v_z u_x - v_x u_z)\vec{Y} + (v_x u_y - v_y u_x)\vec{Z}. \tag{B.9}$$

The vector and every of its components may be a function of the three coordinates (x, y, z). Derivatives of vectors are often described using

$$\nabla = \left(\frac{\partial}{\partial x}, \frac{\partial}{\partial y}, \frac{\partial}{\partial z}\right) = \vec{X}\frac{\partial}{\partial x} + \vec{Y}\frac{\partial}{\partial y} + \vec{Z}\frac{\partial}{\partial z} \tag{B.10}$$

with the divergency

$$\nabla \cdot \vec{v} = \operatorname{div} v = \frac{\partial v_x}{\partial x} + \frac{\partial v_y}{\partial y} + \frac{\partial v_z}{\partial z}, \tag{B.11}$$

and the curl

$$\nabla \times \vec{v} = \operatorname{rot} \vec{v} = \operatorname{curl} \vec{v}$$
$$= \left(\frac{\partial v_z}{\partial y} - \frac{\partial v_y}{\partial z}\right)\vec{X} + \left(\frac{\partial v_x}{\partial z} - \frac{\partial v_z}{\partial x}\right)\vec{Y} + \left(\frac{\partial v_y}{\partial x} - \frac{\partial v_x}{\partial y}\right)\vec{Z} \tag{B.12}$$

Vectors may be generated by the slope of a scalar function $f(x, y, z)$

$$\nabla f = \operatorname{grad} f = \frac{\partial f}{\partial x}\vec{X} + \frac{\partial f}{\partial y}\vec{Y} + \frac{\partial f}{\partial z}\vec{Z} \tag{B.13}$$

The Laplacian Δ is defined by

$$\Delta f = \nabla \cdot \nabla f = \nabla^2 f = \text{div} \cdot \text{grad } f = \frac{\partial^2 f}{\partial x^2} + \frac{\partial^2 f}{\partial y^2} + \frac{\partial^2 f}{\partial z^2}. \qquad \text{(B.14)}$$

A 3×3 matrix (tensor) $\hat{m}$ is given by

$$\hat{m} = \begin{pmatrix} m_{xx} & m_{xy} & m_{xz} \\ m_{yx} & m_{yy} & m_{yz} \\ m_{zx} & m_{zy} & m_{zz} \end{pmatrix}, \qquad \text{(B.15)}$$

it's multiplication by a vector $\vec{u}$ results in another vector $\vec{v} = \hat{m}\,\vec{u}$ with the vector components

$$v_x = m_{xx}u_x + m_{xy}u_y + m_{xz}u_z \qquad \text{(B.16)}$$
$$v_y = m_{yx}u_x + m_{yy}u_y + m_{yz}u_z \qquad \text{(B.17)}$$
$$v_z = m_{zx}u_x + m_{zy}u_y + m_{zz}u_z. \qquad \text{(B.18)}$$

More general, the matrix elements are often given as m_{ij} with $i = 1 \ldots n$ and $j = 1 \ldots m$ for a matrix with n rows and m columns. The determinant of a rectangular matrix $(n = m)$ is written as

$$\det \hat{m} = \begin{vmatrix} m_{11} & m_{12} \\ m_{21} & m_{22} \end{vmatrix} = m_{11}m_{22} - m_{12}m_{21} \qquad \text{(B.19)}$$

for $n = m = 2$, or, for $n = m = 3$, as

$$\det \hat{m} = \begin{vmatrix} m_{11} & m_{12} & m_{13} \\ m_{21} & m_{22} & m_{23} \\ m_{31} & m_{32} & m_{33} \end{vmatrix}$$
$$= m_{11}m_{22}m_{33} - m_{11}m_{32}m_{23} + m_{21}m_{32}m_{13}$$
$$- m_{21}m_{12}m_{33} + m_{31}m_{22}m_{13} - m_{31}m_{22}m_{31}. \qquad \text{(B.20)}$$

B.3   Complex Numbers

A complex number has the form

$$\tilde{c} = a + ib \qquad \text{(B.21)}$$

where a and b are real numbers and $i = \sqrt{-1}$ is the imaginary unit. We call $a = \text{Re}\{\tilde{c}\}$ real part and $b = \text{Im}\{\tilde{c}\}$ imaginary part of $\tilde{c}$ (Cartesian form). Complex numbers can also be writen in polar form,

$$\tilde{c} = r \exp(i\phi) = r \{\cos(\phi) + i \sin(\phi)\}, \qquad \text{(B.22)}$$

where r, called the modulus, is given by

$$r = |\tilde{c}| = \sqrt{a^2 + b^2}, \tag{B.23}$$

and ϕ is called the argument with

$$\phi = \arg \tilde{c} = \arctan \left(\frac{b}{a} \right). \tag{B.24}$$

The complex conjugate of $\tilde{c}$ is defined as

$$\tilde{c}^* = a - ib = r \exp(-i\phi). \tag{B.25}$$

The following rules apply to operations between complex numbers

$$\tilde{c}_1 \pm \tilde{c}_2 = (a_1 \pm a_2) + i(b_1 \pm b_2) \tag{B.26}$$

$$\tilde{c}_1 \tilde{c}_2 = (a_1 a_2 - b_1 b_2) + i(a_1 b_2 + a_2 b_1) = r_1 r_2 e^{i(\phi_1 + \phi_2)} \tag{B.27}$$

$$\frac{\tilde{c}_1}{\tilde{c}_2} = \frac{(a_1 a_2 + b_1 b_2) + i(a_2 b_1 - a_1 b_2)}{a_2^2 + b_2^2} = \frac{r_1}{r_2} e^{i(\phi_1 - \phi_2)}. \tag{B.28}$$

Note that complex numbers are artificial constructions for mathematical convenience. Any real physical quantity is represented by a real number which can be the real part a or the imaginary part b of a complex number (cf. Section 4.2).

B.4 Bessel Functions

Bessel functions are solutions to the Bessel differential equation for the function $f(x)$

$$x^2 \frac{d^2 f}{dx^2} + x \frac{df}{dx} + (x^2 - n^2)f = 0, \tag{B.29}$$

which appears, for instance, in the analysis of cylindrical optical waveguides [256]. Bessel functions can be used to describe the electromagnetic fields of waveguide modes. For integer n, they are given by

$$J_n(x) = \sum_{m=1}^{\infty} \frac{(-1)^m \left(\frac{x}{2} \right)^{n+2m}}{m!(n+m)!} \tag{B.30}$$

with integer m and

$$m! = 1 \times 2 \times \cdots \times (m-1) \times sm. \tag{B.31}$$

Some Bessel functions are plotted in Fig. B.2 (cf. Section 8.6).

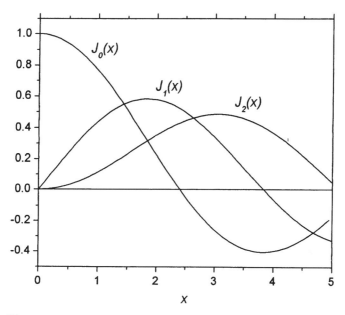

Figure B.2: Bessel functions of zeroth, first, and second order.

B.5 Fourier Transformation

The Fourier integral transformation of the function $f(x)$ is commonly given by

$$F(k) = \frac{1}{\sqrt{2\pi}} \int_{-\infty}^{\infty} dx\, f(x)\, \exp(ikx), \qquad (B.32)$$

and its inverse by

$$f(x) = \frac{1}{\sqrt{2\pi}} \int_{-\infty}^{\infty} dk\, F(k)\, \exp(-ikx). \qquad (B.33)$$

Both transformations are often used to translate functions from real space (x, y, z) into $\vec{k}$ space and vice versa (cf. Section 4.13) or between time and frequency domain.

Appendix C

Symbols and Abbreviations

This list shows symbols of physical parameters frequently used in the book and it gives the introducing section:

a_0	Crystal lattice constant (Section 2.1.2)
a_c	Deformation potential for conduction band (Section 2.2.2)
a_s	Substrate lattice constant (Section 2.2.2)
a_v	Deformation potential for valence band (Section 2.2.2)
b	Shear deformation potential (Section 2.2.2)
B	Spontaneous emission coefficient (Section 3.7)
c	Light velocity (Section 4.2)
C	Net Auger recombination coefficient (Section 3.7)
C_n	Auger recombination coefficient for the conduction band (Section 3.7)
C_v	Auger recombination coefficient for the valence band (Section 3.7)
d	Layer thickness
D_c	Density of states in the conduction band (Section 1.2)
D_n	Electron diffusion coefficient (Section 3.1)
D_p	Hole diffusion coefficient (Section 3.1)
D_r	Reduced density of states (Section 5.1)
D_v	Density of states in the valance band (Section 1.2)
E	Electron energy (Section 1.2)
$\vec{E}$	Electrical field vector of electromagnetic waves (Section 4.1)
$\vec{\mathrm{E}}$	Time-independent electrical field amplitude (Section 4.1)
E_c, E_c^0	Conduction band edge energy (Section 1.2)
E_g	Band gap energy (Section 1.1)
E_{hh}	Heavy-hole energy (Section 2.2.1)
E_{lh}	Light-hole energy (Section 2.2.1)
E_v, E_v^0	Valence band edge energy (Sections 1.2, 2.5)
E_F	Fermi energy (Section 1.2)
E_{Fn}	Quasi-Fermi energy for electrons (Section 3.2)
E_{Fp}	Quasi-Fermi energy for holes (Section 3.2)

f	Fermi distribution function for electrons (Section 1.2)
$\vec{F}$	Electrostatic field vector (Section 3.1)
$F_{1/2}$	Fermi integral (Section 1.3)
G	Free carrier generation rate (Section 3.1)
h	Planck's constant (Section A.1)
$\hbar$	Reduced Planck's constant $\hbar = h/2\pi$ (Section A.1)
$\vec{H}$	Magnetic field vector of electromagnetic waves (Section 4.1)
$\vec{H}$	Time-independent magnetic field amplitude (Section 4.1)
i	Imaginary unit $i = \sqrt{-1}$ (Section B.3)
I	Electrical current
I_{opt}	Optical intensity (Section 4.4)
I_{th}	Threshold current for lasing (Section 7.1)
$\vec{j}_n$	Electron current density (Section 3.1)
$\vec{j}_p$	Hole current density (Section 3.1)
$\vec{k}$	Wave vector (Sections 2.1, 4.4)
k_{a}	Optical extinction ratio (Section 4.2)
k_{B}	Boltzmann constant (Section A.1)
L	Device length
L	Characteristic point in the Brillouin zone (Section 2.1.2)
m_{c}	Electron effective mass (Section 1.2)
m_{hh}	Heavy hole effective mass (Section 2.1.2)
m_{lh}	Light hole effective mass (Section 2.1.2)
m_{v}	Hole effective mass (Section 1.2)
m_0	Free electron mass (Section A.1)
M_{b}	Bulk momentum matrix element (Section 5.1.1)
n	Electron concentration (Section 1.2)
n_0	Equilibrium electron concentration (Section 3.5.2)
N	Average carrier concentration (Section 2.1.3)
n_{A}	Concentration of ionized acceptors (negatively charged) (Section 1.3)
n_{i}	Intrinsic carrier concentration (Section 1.2)
n_{r}	Index of refraction (Section 4.2)
N_{c}	Effective density of states in the conduction band (Section 1.2)
N_{dop}	Doping density (Section 1.3)
N_{v}	Effective density of states in the valence band (Section 1.2)
N_{A}	Acceptor atom concentration (Section 1.3)
N_{D}	Donor atom concentration (Section 1.3)
p	Hole concentration (Section 1.2)

p_0	Equilibrium hole concentration (Section 3.5.2)
p_D	Concentration of ionized donors (positively charged) (Section 1.3)
P_{heat}	Heat power (Section 6.3)
P_n	Electron thermoelectric power (Section 3.9.1)
P_{opt}	Optical power (Section 4.12)
P_p	Hole thermoelectric power (Section 3.9.1)
q	Elementary charge (Section A.1)
r	Radius (Section B.1)
r_{ij}	Optical field reflection coefficient (Section 4.5)
R	Optical power reflectivity (reflectance) (Section 4.5)
R_{Aug}	Auger recombination rate (Section 3.7)
R_{spon}	Spontaneous photon emission rate (Section 3.7)
R_{SRH}	Shockley–Read–Hall recombination rate (Section 3.7)
R_{stim}	Stimulated photon emission rate (Section 3.7)
R_{th}	Thermal resistance (Section 6.3)
t	Time
t_{ij}	Optical field transmission coefficient (Section 4.5)
T	Optical power transmission coefficient (transmittance) (Section 4.5)
T	Temperature
T_L	Crystal lattice temperature (Section 6.1)
T_n	Electron temperature (Section 3.9.1)
T_p	Hole temperature (Section 3.9.1)
v	Carrier velocity
V	Applied voltage; bias
x	Coordinate in the Cartesian system (Section B.1)
X	Characteristic Point in the Brillouin zone (Section 2.1.2)
y	Coordinate in the Cartesian system (Section B.1)
z	Coordinate in the Cartesian system (Section B.1)
α_i	Internal modal absorption coefficient (Section 7.1)
α_o	Optical absorption coefficient (Section 4.2)
β_m	Propagation constant of waveguide mode m (Section 4.8)
Γ	Center of the Brillouin zone (Section 2.1.2)
Γ_o	Optical confinement factor (Section 7.1)
Δ_0	Split-off energy (Section 2.1.2)
ε_0	Vacuum electrical permittivity (Section 1.1)
ε_{ij}	Lattice strain (Section 2.2.2)
ε_{st}	Static dielectric constant (Section 4.1)
ε_{opt}	Optical (high-frequency) dielectric constant (Section 4.1)
η	Quantum efficiency (Section 7.1)

θ	Interface bounce angle of optical wave (Section 4.8)
ϑ	Angle of incidence of optical wave at interface (Section 4.5)
Θ	Angle in spherical coordinate system (Section B.1)
κL	Thermal conductivity of crystal lattice (Section 6.1)
λ	Optical wavelength (Section 4.2)
λ_g	Optical wavelength equivalent to band gap energy E_g (Section 1.1)
μ	Electron or hole mobility (Section 3.6)
μ_0	Vacuum magnetic permeability (Section 4.1)
ν	Optical frequency (Section 1.1)
π	$= 3.1416$, geometric constant (Section A.1)
σ	Electrical conductivity (Section 1.1)
τ_n	Electron recombination lifetime (Section 3.7)
τ_p	Hole recombination lifetime (Section 3.7)
τ_s	Intraband carrier scattering lifetime (Section 5.1.2)
φ	Electrostatic potential (Section 3.2)
φ_r	Reflection phase shift (Section 4.5)
ω	Angular optical frequency $\omega = 2\pi\nu$ (Section 1.1)

The following abbreviations are used in the book:

1D	One-dimensional
2D	Two-dimensional
3D	Three-dimensional
APSYS	Simulation software by Crosslight Software, Inc.
BPM	Beam propagation method
CH	Crystal-field split-off hole
DBR	Distributed Bragg reflector
DFB	Distributed feedback
DOS	Density of States
fcc	Face-centered cubic
FWHM	Full-width half-maximum
HH	Heavy hole
IV	Current vs voltage
LASTIP	Simulation software by Crosslight Software, Inc.
LED	Light-emitting diode
LI	Light vs current
LH	Light hole
LHS	Left-hand side
MQW	Multiquantum well
PICS3D	Simulation software by Crosslight Software, Inc.

QCSE	Quantum-confined Stark effect
QW	Quantum well
PL	Photoluminescence
RHS	Right-hand side
SCL	Separate confinement layer
SO	Spin-orbit split-off
SRH	Shockley–Read–Hall
TE	Transverse electric
TEM	Transverse electromagnetic
TM	Transverse magnetic
VCSEL	Vertical-cavity surface-emitting laser

Bibliography

[1] O. Madelung, ed., *Semiconductors - Basic Data*. Berlin: Springer-Verlag, 1996.

[2] K. W. Boer, *Survey of Semiconductor Physics*, vol. I. New York: Van Nostrand Reinhold, 1990.

[3] L. I. Berger, *Semiconductor Materials*. Boca Raton, FL: CRC Press, 1997.

[4] M. E. Levinshtein, S. L. Rumyantsev, and M. S. Shur, eds., *Handbook Series on Semiconductor Parameters*, vol. 1. Singapore: World Scientific, 1996.

[5] V. Swaminathan, Properties of InP and related materials, in *Indium Phosphide and Related Materials* (A. Katz, ed.), pp. 1–44, Boston: Artec House, 1992.

[6] S. Jain, M. Willander, and R. V. Overstraeten, *Compound Semiconductors, Strained Layers, and Devices*. Dordrecht: Kluwer, 2000.

[7] W. Nakwaski, Effective masses of electrons and heavy holes in GaAs, InAs, AlAs and their ternary compounds, *Physica B*, vol. 210, pp. 1–25, 1995.

[8] D. Bednarczyk and J. Bednarczyk, The approximation of the Fermi-Dirac integral $f_{1/2}(\eta)$, *Phys. Lett.*, vol. 64A, pp. 409–410, 1978.

[9] S. Tiwari, *Compound Semiconductor Device Physics*. San Diego: Academic Press, 1992.

[10] W. W. Chow and S. W. Koch, *Semiconductor-Laser Fundamentals*. Berlin: Springer-Verlag, 1999.

[11] Courtesy of W. Frensley, University of Texas at Dallas, available at http://www.utdallas.edu/~frensley.

[12] N. W. Ashcroft and N. D. Mermin, *Solid State Physics*. Forth Worth: Harcourt Brace College, 1976.

[13] I. Vurgaftman, J. R. Meyer, and L. R. Ram-Mohan, Band parameters for III-V compound semiconductors and their alloys, *J. Appl. Phys.*, vol. 89, pp. 5815–5875, 2001.

[14] J. C. Hensel, H. Hasegawa, and M. Nakayama, Cyclotron resonance in uniaxially stressed silicon, *Phys. Rev. A*, vol. 138, pp. 225–238, 1965.

[15] Y. P. Varshni, Temperature dependence of the energy gap in semiconductors, *Physica*, vol. 34, pp. 149–154, 1967.

[16] R. Pässler, Parameter sets due to fittings of the temperature dependencies of fundamental bandgaps in semiconductors, *phys. stat. sol. (b)*, vol. 216, pp. 975–1007, 1999.

[17] R. Zimmermann, *Many-particle theory of highly excited semiconductors.* Leipzig: Teubner Verlagsgesellschaft, 1988.

[18] D. B. M. Klaassen, J. W. Slotboom, and H. C. de Graaff, Unified apparent bandgap narrowing in n- and p-type silicon, *Solid State Electron.*, vol. 35, pp. 125–129, 1992.

[19] V. Palankovski, G. Kaiblinger-Grujin, and S. Selberherr, Study of dopant-dependent band gap narrowing in compound semiconductor devices, *Mater. Sci. Eng. B*, vol. 66, pp. 46–49, 1999.

[20] S. L. Chuang, *Physics of Optoelectronic Devices.* New York: Wiley, 1995.

[21] S. Chuang, Efficient band-structure calculation of strained quantum-wells, *Phys. Rev. B*, vol. 43, pp. 9649– 9661, 1991.

[22] G. L. Bir and G. E. Pikus, eds., *Symmetry and Strain Induced Effects in Semiconductors.* New York: Wiley, 1974.

[23] C. Van De Walle, Band lineups and deformation potentials in the model-solid theory, *Phys. Rev. B*, vol. 39, pp. 1871–1883, 1989.

[24] J. Luttinger and W. Kohn, Motion of electrons and holes in perturbed periodic fields, *Phys. Rev.*, vol. 97, pp. 869– 883, 1955.

[25] C. Y. P. Chao and S. L. Chuang, Spin-orbit-coupling effects on the valence band structure of strained semiconductor quantum wells, *Phys. Rev. B*, vol. 46, pp. 4110–4122, 1992.

[26] E. O. Kane, Energy band theory, in *Handbook on Semiconductors* (T. S. Moss, ed.), pp. 193–217, New York: North-Holland, 1982.

[27] P. Enders, A. Bärwolff, M. Woerner, and D. Suisky, kp theory of energy bands, wave functions, and optical selection rules in strained tetrahedral semiconductors, *Phys. Rev. B*, vol. 51, pp. 16695–16704, 1995.

[28] J. P. Loehr, *Physics of Strained Quantum Well Lasers*. Boston: Kluwer, 1999.

[29] S. L. Chuang and C. S. Chang, A band-structure model of strained quantum-well wurtzite semiconductors, *Semicond. Sci. Technol.*, vol. 12, pp. 252–263, 1997.

[30] S. L. Chuang and C. S. Chang, k.p method for strained wurtzite semiconductors, *Phys. Rev. B*, vol. 54, pp. 2491–2504, 1996.

[31] M. Kumagai, S. L. Chuang, and H. Ando, Analytical solutions of the block-diagonalized hamiltonian for strained wurtzite semiconductors, *Phys. Rev. B*, vol. 57, pp. 15303–15314, 1998.

[32] S. Strite and H. Morkoc, GaN, AlN, and InN: A review, *J. Vac. Sci. Technol.*, vol. 10, no. 4, pp. 1237–1266, 1992.

[33] S.-H. Wei and A. Zunger, Valence band spittings and band offsets of AlN, GaN, and InN, *Appl. Phys. Lett.*, vol. 69, pp. 2719–2721, 1996.

[34] S. L. Chuang, Optical gain of strained wurtzite GaN quantum well lasers, *IEEE J. Quantum Electron.*, vol. 32, no. 10, pp. 1791–1799, 1996.

[35] Y. C. Yeo, T. C. Chong, and M. F. Li, Electronic band structures and effective-mass parameters of wurtzite GaN and InN, *J. Appl. Phys.*, vol. 83, pp. 1429–1436, 1998.

[36] C. Wetzel, T. Takeuchi, S. Yamaguchi, H. Katoh, H. Amano, and I. Akasaki, Optical band gap in GaInN on GaN by photoreflection spectroscopy, *Appl. Phys. Lett.*, vol. 73, pp. 1994–1996, 1998.

[37] W. W. Chow, M. Hagerott-Crawford, A. Girndt, and S. W. Koch, Threshold conditions for an ultraviolet wavelength GaN quantum-well laser, *IEEE J. Select. Topics Quantum Electron.*, vol. 4, pp. 514–519, 1998.

[38] M. E. Levinshtein, S. L. Rumyantsev, and M. S. Shur, eds., *Properties of advanced semiconductor materials*. New York: Wiley, 2001.

[39] L. A. Coldren and S. W. Corzine, *Diode Lasers and Photonic Integrated Physics*. New York: Wiley, 1995.

[40] K. Nakamura, A. Shimizu, M. Koshiba, and K. Hayata, Finite-element analysis of quantum wells of arbitrary semiconductors with arbitrary potential profiles, *IEEE J. Quantum Electron.*, vol. 25, pp. 889–895, 1989.

[41] D. Ahn, S. L. Chuang, and Y. C. Chang, Valence-band mixing effects on the gain and the refractive index change of quantum well lasers, *J. Appl. Phys.*, vol. 64, pp. 4056–4064, 1988.

[42] M. P. C. M. Krijn, Heterojunction band offsets and effective masses in III-V quaternary alloys, *Semicond. Sci. Technol.*, vol. 6, pp. 27–31, 1991.

[43] M. Guden and J. Piprek, Material parameters of quaternary III-V semi-conductors for multilayer mirrors at 1.55 micron wavelength, *Modelling Simulation Mater. Sci. Eng.*, vol. 4, no. 4, pp. 349–57, 1996.

[44] M. Cardona and N. E. Christensen, Comments on "spectroscopy of excited states in InGaAs-InP single quantum wells grown by chemical-beam epitaxy, *Phys. Rev. B*, vol. 37, pp. 1011–1012, 1988.

[45] C. Van De Walle, Theoretical study of band offsets at semiconductor interfaces, *Phys. Rev. B*, vol. 35, pp. 8154–8165, 1987.

[46] K. W. Boer, *Survey of Semiconductor Physics*, vol. II. New York: Van Nostrand Reinhold, 1992.

[47] Ioffe Institute, St. Petersburg, Russia, Web Archive on Semiconductor Data, available at http://www.ioffe.rssi.ru/SVA/.

[48] W. R. Frensley, Quantum transport, in *Heterostructures and Quantum Devices* (N. G. Einspruch and W. R. Frensley, eds.), pp. 273–303, San Diego: Academic Press, 1994.

[49] R. Lake, G. Klimeck, R. C. Bowen, and D. Jovanovic, Single and multiband modeling of quantum electron transport through layered semiconductor devices, *J. Appl. Phys.*, vol. 81, pp. 7845–7869, 1997.

[50] Y. Ando and T. Itoh, Calculation of transmission tunneling current across arbitrary potential barriers, *J. Appl. Phys.*, vol. 61, pp. 1497–1502, 1987.

[51] A. Messiah, *Quantum Mechanics*. Amsterdam: North-Holland, 1961.

[52] J. Singh, *Physics of Semiconductors and their Heterostructures*. New York: McGraw-Hill, 1993.

[53] A. Schenk, *Physical Models for Silicon Device Simulation*. Berlin: Springer-Verlag, 1998.

[54] E. Burstein and S. Lundqvist, eds., *Tunneling Phenomena in Solids*. New York: Plenum Press, 1969.

[55] H. B. Michaelson, ed., *Handbook of Chemistry and Physics*. Cleveland: Chemical Rubber Co., 1972.

[56] S. Selberherr, *Analysis and simulation of semiconductor devices*. New York: Springer-Verlag, 1984.

[57] S. M. Sze, *Physics of Semiconductor Devices*. New York: Wiley, 2 ed., 1981.

[58] C. M. Snowden, *Semiconductor Device Modelling*. London: Peregrinus, 1988.

[59] Z. M. Li, *Crosslight User Manual*. Burnaby, British Columbia: Crosslight Software, Inc., 2001.

[60] PC1D solar cell simulation software, Key Centre for Photovoltaic Engineering, University of New South Wales, Australia, available at http://www.pv.unsw.edu.au/pc1d/.

[61] MINIMOS field-effect transistor simulation software, Institute for Microelectronics, Technical University, Vienna, Austria, available at http://www.www.iue.tuwien.ac.at/software/.

[62] M. Sotoodeh, A. H. Khalid, and A. A. Rezazadeh, Empirical low-field mobility model for III-V compounds applicable in device simulation, *J. Appl. Phys.*, vol. 87, no. 6, pp. 2890–2900, 2000.

[63] M. Farahmand, C. Garetto, E. Bellotti, K. F. Brennan, M. Goano, E. Ghillino, G. Ghione, J. D. Albrecht, and P. P. Ruden, Monte Carlo simulation of electron transport in the III-nitride wurtzite phase material system: binaries and ternaries, *J. Appl. Phys.*, vol. 48, pp. 535–542, 2001.

[64] SIMBA semiconductor device simulation software, Technical University, Dresden, Germany, available at http://www.iee.et.tu-dresden.de/˜klix/simba/.

[65] D. M. Caughey and R. E. Thomas, Carrier mobilities in silicon empirically related to doping and field, *Proc. IEEE*, vol. 55, pp. 2192–2193, 1967.

[66] C. M. Maziar and M. S. Lundstrom, Caughey-thomas parameters for electron mobility calculations in GaAs, *Electron. Lett.*, vol. 22, pp. 565–566, 1986.

[67] N. D. Arora, J. R. Hauser, and D. J. Roulston, Electron and hole mobilities in silicon as a function of concentration and temperature, *IEEE Trans. Electron Devices*, vol. 29, pp. 292–295, 1982.

[68] R. Quay, C. Moglestue, V. Palankovski, and S. Selberherr, A temperature dependent model for the saturation velocity in semiconductor materials, *Mater. Sci. Semicond. Process.*, vol. 3, pp. 149–155, 2000.

[69] S. Adachi, *Physical Properties of III-V Semiconductor Compounds*. New York: Wiley, 1992.

[70] M. E. Levinshtein, S. L. Rumyantsev, and M. S. Shur, eds., *Handbook Series on Semiconductor Parameters*, vol. 2. Singapore: World Scientific, 1999.

[71] H. Kroemer, The Einstein relation for degenerate carrier concentration, *IEEE Trans. Electron Devices*, vol. 25, p. 850, 1978.

[72] G. E. Shtengel, D. A. Ackermann, P. A. Morton, E. J. Flynn, and M. S. Hybertsen, Impedance-corrected carrier lifetime measurements in semiconductor lasers, *Appl. Phys. Lett.*, vol. 67, pp. 1506–1508, 1995.

[73] M. C. Wang, K. Kash, C. Zah, R. Bhat, and S. L. Chuang, Measurement of nonradiative auger and radiative recombination rates in strained-layer quantum-well systems, *Appl. Phys. Lett.*, vol. 62, pp. 166–168, 1993.

[74] Y. Suematsu and A. R. Adams, eds., *Handbook of Semiconductor Lasers and Photonic Integrated Circuits*. London: Chapman & Hall, 1994.

[75] G. W. Charache, P. F. Baldasaro, L. R. Danielson, D. M. DePoy, M. J. Freeman, C. A. Wang, H. K. Choi, D. Z. Garbuzov, R. U. Martinelli, V. Khalfin, S. Saroop, J. M. Borrego, and R. J. Gutmann, InGaAsSb thermophotovoltaic diode: Physics evaluation, *J. Appl. Phys.*, vol. 85, pp. 2247–2252, 1999.

[76] E. Wintner and E. P. Ippen, Nonlinear carrier dynamics in InGaAsP compounds, *Appl. Phys. Lett.*, vol. 44, pp. 999–1001, 1984.

[77] G. Agrawal and N. Dutta, *Semiconductor Lasers*. New York: van Nostrand Reinhold, 1986.

[78] W. Shockley and J. T. W. Read, Statistics of the recombinations of holes and electrons, *Phys. Rev.*, vol. 87, pp. 835–842, 1952.

[79] R. N. Hall, Electron-hole recombination in germanium, *Phys. Rev.*, vol. 87, p. 387, 1952.

[80] T. P. Pearsall, F. Capasso, R. E. Nahory, M. A. Pollack, and J. R. Chelikowsky, The band structure dependence of impact ionization by hot carriers in semiconductors: GaAs, *Solid State Electron.*, vol. 21, pp. 297–302, 1978.

[81] A. S. Kyuregyan and S. N. Yurkov, Room-temperature avalanche break-down voltages of p-n junctions made of Si, Ge, SiC, GaAs, GaP, and InP, *Sov. Phys.—Semicond.*, vol. 23, pp. 1126–1132, 1989.

[82] R. V. Oversraeten and H. de Man, Measurement of the ionization rates in duffused silicon *pn*-junctions, *Solid State Electron.*, vol. 13, pp. 583–608, 1970.

[83] T. Mikawa, S. Kagawa, T. Kaneda, Y. Toyama, and O. Mikami, Crystal orientation dependence of ionization rates in germanium, *Appl. Phys. Lett.*, vol. 37, pp. 387–389, 1980.

[84] P. Yuan, K. A. Anselm, C. Hu, H. Nie, C. Lenox, A. L. Holmes, B. G. Streetman, J. C. Campbell, and R. J. McIntyre, A new look at impact ion-ization part II: Gain and noise in short avalanche photodiodes, *IEEE Trans. Electron Devices*, vol. 46, pp. 1632–1638, 1999.

[85] C. A. Armiento and S. H. Grove, Impact ionization in (100)-, (110)-, and (111)-oriented InP avalanche photodiodes, *Appl. Phys. Lett.*, vol. 43, pp. 198–200, 1983.

[86] G. A. Baraff, Distribution functions and ionization rates for hot electrons in semiconductors, *Phys. Rev.*, vol. 128, pp. 2507–2517, 1962.

[87] C. R. Crowell and S. M. Sze, Temperature dependence of avalanche multi-plication in semiconductors, *Appl. Phys. Lett.*, vol. 9, pp. 242–244, 1966.

[88] A. P. Dmitriev, M. P. Mikhailova, and I. N. Yassievich, Impact ionization in $A^{III}B^{V}$ semiconductors in high electric fields *phys. stat. sol. (b)*, vol. 140, pp. 9–37, 1987.

[89] K. Hess, *Advanced Theory of Semiconductor Devices*. New York: IEEE Press, 2000.

[90] R. Adar, Spatial integration of direct band-to-band tunneling currents in general device structures, *IEEE Trans. Electron Devices*, vol. 39, pp. 976–981, 1992.

[91] G. A. M. Hurkx, D. B. M. Klaassen, and M. P. G. Knuvers, A new recombina-tion model for device simulation including tunneling, *IEEE Trans. Electron Devices*, vol. 39, pp. 331–338, 1992.

[92] R. Stratton, Diffusion of hot and cold electrons in semiconductor barriers, *Phys. Rev.*, vol. 126, pp. 2002–2014, 1962.

[93] K. Bløtekjær, Transport equations for electrons in two-valley semiconduc-tors, *IEEE Trans. Electron Devices*, vol. 17, pp. 38–47, 1970.

[94] Y. Apanovich, E. Lyumkis, B. Polsky, A. Shur, and P. Blakey, Steady-state and transient analysis of submicron devices using energy balance and simplified hydrodynamic models, *IEEE Trans. Computer Aided Design*, vol. 13, pp. 702–710, 1994.

[95] K. Hess, *Monte Carlo Device Simulation: Full Band and Beyond*. Boston: Kluwer, 1991.

[96] A. Forghieri, R. Guerrieri, P. Ciampolini, A. Gnudi, M. Rudan, and G. Baccarani, A new discretization strategy of the semiconductor equations comprising momentum and energy balance, *IEEE Trans. Computer Aided Design*, vol. 7, pp. 231–242, 1988.

[97] D. Chen, E. Sangiorgi, M. R. Pinto, E. C. Kan, U. Ravaioli, and R. W. Dutton, An improved energy-transport model including nonparabolicity and non-maxwellian distribution effects, *IEEE Trans. Computer Aided Design*, vol. 13, pp. 26–28, 1992.

[98] J. F. Nye, *Physical Properties of Crystals*. Oxford: Clarendon Press, 1985.

[99] E. D. Palik, ed., *Handbook of Optical Constants of Solids*. San Diego: Academic Press, 1998.

[100] F. Fiedler and A. Schlachetzki, Optical parameters of InP-based waveguides, *Solid State Electron.*, vol. 30, no. 1, pp. 73–83, 1987.

[101] C. Henry, R. Logan, F. R. Merritt, and J. P. Luongo, The effect of intervalence band absorption on the thermal behaviour of InGaAsP lasers, *IEEE J. Quantum Electron.*, vol. 19, no. 6, pp. 947–952, 1983.

[102] J. Taylor and V. Tolstikhin, Intervalence band absorption in InP and related materials for optoelectronic device modeling, *J. Appl. Phys.*, vol. 87, pp. 1054–1059, 2000.

[103] P. Y. Yu and M. Cardona, *Fundamentals of Semiconductors*. Berlin: Springer-Verlag, 1996.

[104] G. Ghosh, Refractive index, in *Handbook of Optical Constants of Solids* (D. Palik and G. Ghosh, eds.), pp. 5–110, San Diego: Academic Press, 1998.

[105] M. J. Mondry, D. I. Babic, J. E. Bowers, and L. Coldren, Refractive indexes of (Al,Ga,In)As epilayers on InP for optoelectronic applications, *IEEE Photon. Technol. Lett.*, vol. 4, no. 6, pp. 627–629, 1992.

[106] S. Adachi, H. Kato, A. Moki, and K. Ohtsuka, Refractive index of AlGaInP quaternary alloys, *J. Appl. Phys.*, vol. 75, no. 1, pp. 478–80, 1994.

[107] S. H. Wemple and J. DiDomenico, M, Behavior of the electronic dielectric constant in covalent and ionic materials, *Phys. Rev. B*, vol. 3, no. 4, pp. 1338–1351, 1971.

[108] T. Takagi, Dispersion parameters of the refractive index in III-V compound semiconductors, *Japan. J. Appl. Phys., Part 1*, vol. 21, no. 3, pp. L167–L169, 1982.

[109] K. Utaka, K.-I. Kobayashi, and Y. Suematsu, Lasing characteristics of 1.5–1.6 μm GaInAsP/InP integrated twin-guide lasers with first-order distributed Bragg reflectors, *IEEE J. Quantum Electron.*, vol. 17, pp. 651–658, 1981.

[110] M. A. Afromowitz, Refractive index of GaAlAs, *Solid State Comm.*, vol. 15, pp. 59–63, 1974.

[111] J. Buus and M. J. Adams, Phase and group indices for double heterostructure lasers, *IEE J. Solid State Electron. Devices*, vol. 3, pp. 189–195, 1979.

[112] T. Peng and J. Piprek, Refractive index of AlGaInN alloys, *Electron. Lett.*, vol. 32, pp. 2285–2286, 1996.

[113] S. Adachi, Optical dispersion relations for Si and Ge, *J. Appl. Phys.*, vol. 66, no. 7, pp. 3224–3231, 1989.

[114] S. Adachi, Optical properties of AlGaAs alloys, *Phys. Rev. B*, vol. 38, no. 17, pp. 12345–12352, 1988.

[115] S. Adachi, Refractive indices of III-V compounds: key properties of InGaAsP relevant to device design, *J. Appl. Phys.*, vol. 53, no. 8, pp. 5863–5869, 1982.

[116] S. Adachi, Band gaps and refractive indices of AlGaAsSb, GaInAsSb, and InPAsSb: key properties for a variety of the 2-4 micron optoelectronic device applications, *J. Appl. Phys.*, vol. 61, no. 10, pp. 4869–4876, 1987.

[117] H. Kato, S. Adachi, H. Nakanishi, and K. Ohtsuka, Optical properties of AlGaInP quaternary alloys, *Japan. J. Appl. Phys., Part 1*, vol. 33, no. 1A, pp. 186–192, 1994.

[118] S. Adachi and T. Kimura, Optical constants of ZnCdTe ternary alloys: experiment and modeling, *Japan. J. Appl. Phys., Part 1*, vol. 32, no. 8, pp. 3496–3501, 1993.

[119] J. Piprek, T. Peng, G. Qui, and J. O. Olowolafe, Energy gap bowing and refractive index spectrum of AlInN and AlGaInN, in *IEEE International*

Symposium on Compound Semiconductors (M. Melloch and M. A. Reed, eds.), (Bristol), pp. 227–230, Institute of Physics, 1997.

[120] G. M. Laws, E. C. Larkins, I. Harrison, C. Molloy, and D. Somerford, Improved refractive index formulas for the AlGaN and InGaN alloys, *J. Appl. Phys.*, vol. 89, pp. 1108–1115, 2001.

[121] B. Bennett, R. Soref, and J. Del Alamo, Carrier-induced change in refractive index of InP, GaAs and InGaAsP, *IEEE J. Quantum Electron.*, vol. 26, no. 1, pp. 113–122, 1990.

[122] P. Yeh, *Optical Waves in Layered Media*. New York: Wiley, 1988.

[123] R. Scarmozzino, A. Gopinath, R. Pregla, and S. Helfert, Numerical techniques for modeling guided-wave photonic devices, *IEEE J. Select. Topics Quantum Electron.*, vol. 6, no. 1, pp. 150–162, 2000.

[124] K. S. Chiang, Review of numerical and approximate methods for the modal analysis of general optical dielectric waveguides, *Opt. Quantum Electron.*, vol. 26, pp. S113–S134, 1994.

[125] G. Guekos, ed., *Photonic Devices for Telecommunications*. Berlin: Springer-Verlag, 1999.

[126] B. E. A. Saleh and M. C. Teich, *Fundamentals of Photonics*. New York: Wiley, 1991.

[127] H. Kressel and J. K. Butler, *Semiconductor Lasers and Heterojunction LEDs*. New York: Academic Press, 1977.

[128] K. J. Ebeling, *Integrated Optoelectronics*. Berlin: Springer-Verlag, 1993.

[129] J.-P. Berenger, A perfectly matched layer for the absorption of electromagnetic waves, *J. Comput. Phys.*, vol. 114, pp. 185–200, 1994.

[130] E. A. J. Marcatili, Dielectric rectangular waveguide and directional coupler for integrated optics, *Bell Syst. Tech. J.*, vol. 48, pp. 2071–2102, 1969.

[131] D. Marcuse, *Theory of Dielectric Optical Waveguides*. Boston: Academic Press, 1991.

[132] T. Ikegami, Reflectivity of mode at facet and oscillation mode in double-heterostructure injection lasers, *IEEE J. Quantum Electron.*, vol. 8, pp. 470–476, 1972.

[133] L. Lewin, A method for the calculation of the radiation-pattern and mode-conversion properties of a solid-state heterojunction laser, *IEEE Trans. Microwave Theory Techn.*, vol. 23, pp. 576–585, 1975.

[134] C. Herzinger, C. Lu, T. DeTemple, and W.C.Chew, The semiconductor waveguide facet reflectivity problem, *IEEE J. Quantum Electron.*, vol. 29, pp. 2273–2281, 1993.

[135] J. Buus, Analytical approximation for the reflectivity of dh lasers, *IEEE J. Quantum Electron.*, vol. 17, pp. 2256–2257, 1981.

[136] H. Wenzel, *Modellierung von Quasi-Indexgeführten Halbleiterlasern*. PhD thesis, Humboldt University Berlin, 1991.

[137] P. C. Kendall, D. A. Roberts, P. N. Robson, M. J. Adams, and M. J. Robertson, Semiconductor laser facet reflectivities using free-space radiation modes, *IEE Proc., Part J: Optoelectron.*, vol. 140, pp. 49–55, 1993.

[138] M. Reed, T. M. Benson, P. C. Kendall, and P. Sewell, Antireflection-coated angled facet design, *IEE Proc., Part J: Optoelectron.*, vol. 143, pp. 214–220, 1996.

[139] J. Carroll, J. Whiteaway, and D. Plump, *Distributed Feedback Semiconductor Lasers*. Bellingham, WA: SPIE Optical Engineering Press, 1998.

[140] G. Morthier and P. Vankwikelberge, *Handbook of Distributed Feedback Laser Diodes*. Norwood, MA: Artech House, 1997.

[141] W. D. Herzog, M. S. Unlu, B. B. Goldberg, G. H. Rhodes, and C. Harder, Beam divergence and waist measurements of laser diodes by near-field scanning optical microscopy, *Appl. Phys. Lett.*, vol. 70, no. 6, pp. 688–690, 1976.

[142] A. E. Siegman, *Lasers*. Mill Valley: University Science Books, 1986.

[143] W. T. Silfvast, *Laser Fundamantals*. Cambridge, UK: Cambridge Univ. Press, 1996.

[144] J. Buus, Beamwidth for asymmetric and mulitlayer semiconductor laser structures, *IEEE J. Quantum Electron.*, vol. 17, pp. 732–736, 1981.

[145] T. Lau and J. M. Ballantyne, Two dimensional analysis of a dielectric waveguide mirror, *IEEE J. Lightwave Technol.*, vol. 48, pp. 551–558, 1997.

[146] R. Yan, S. Corzine, L. Coldren, and I. Suemune, Corrections to the expression of gain in GaAs, *IEEE J. Quantum Electron.*, vol. 26, pp. 213–216, 1990.

[147] S. Chinn, P. Zory, and A. Reisinger, A model for GRIN-SCH-SQW diode lasers, *IEEE J. Quantum Electron.*, vol. 24, pp. 2191–2214, 1988.

[148] P. Landsberg, Electronic interaction effects on recombination spectra, *Phys. Stat. Sol. (b)*, vol. 15, pp. 623–626, 1966.

[149] R. Martin and H. Stormer, On the low energy tail of the electron-hole drop recombination spectrum, *Solid State Comm.*, vol. 22, pp. 523–526, 1977.

[150] E. Zielinski, H. Schweizer, S. Hausser, R. Stuber, M. H. Pilkuhn, and G. Weimann, Systematics of laser operation in GaAs/AlGaAs multi-quantum well heterostructures, *IEEE J. Quantum Electron.*, vol. 23, pp. 969–976, 1987.

[151] E. Zielinski, F. Keppler, S. Hausser, M. H. Pilkuhn, R. Sauer, and W. Tsang, Optical gain and loss processes in GaInAs/InP-MQW laser structure, *IEEE J. Quantum Electron.*, vol. 25, pp. 1407–1416, 1989.

[152] M. Asada, Intraband relaxation time in quantum-well lasers, *IEEE J. Quantum Electron.*, vol. 25, no. 9, pp. 2019–2026, 1993.

[153] M. Asada, Intraband relaxation effect on optical spectra, in *Quantum Well Lasers* (P. S. Zory, ed.), pp. 97–130, San Diego: Academic Press, 1993.

[154] *IEEE Journal of Quantum Electronics*, Special Issue on Strained-Layer Optoelectronic Materials and Devices, February 1994.

[155] C. Ohler, J. Moers, A. Forster, and H. Luth, Strain dependence of the valence-band offset in arsenide compound heterojunctions determined by photoelectron spectroscopy, *J. Vac. Sci. Technol.*, vol. B13, pp. 1728–1735, 1995.

[156] C.-F. Hsu, P. Zory, C.-H. Wu, and M. Emanuel, Coulomb enhancement in InGaAs-GaAs quantum-well lasers, *IEEE J. Select. Topics Quantum Electron.*, vol. 3, pp. 158–165, 1997.

[157] D. A. B. Miller, D. S. Chemla, T. C. Damen, A. C. Gossard, W. Wiegmann, T. H. Wood, and C. A. Burrus, Electric field dependence of optical absorption near the band gap of quantum well structures, *Phys. Rev. B*, vol. 32, pp. 1043–1059, 1985.

[158] C. Kittel, *Introduction to Solid State Physics*. New York: Wiley, 1996.

[159] G. Wachutka, Rigorous thermodynamic treatment of heat generation and conduction in semiconductor modeling, *IEEE Trans. Computer Aided Design*, vol. 9, pp. 1141–1149, 1990.

[160] W. Nakwaski, Thermal conductivity of binary, ternary and quaternary III-V compounds, *J. Appl. Phys.*, vol. 64, pp. 159–166, 1988.

[161] B. Abeles, Lattice thermal conductivity of disordered semiconductor alloys at high temperatures, *Phys. Rev.*, vol. 131, pp. 1906–1911, 1963.

[162] P. Abraham, J. Piprek, S. DenBaars, and J. E. Bowers, Study of temperature effects on loss mechanisms in 1.55 μm laser diodes with InGaP electron stopper layer, *Semicond. Sci. Technol.*, vol. 14, pp. 419–424, 1999.

[163] J. Piprek, P. Abraham, and J. E. Bowers, Carrier nonuniformity effects on the internal efficiency of multiquantum-well lasers, *Appl. Phys. Lett.*, vol. 74, no. 4, pp. 489–491, 1999.

[164] J. Piprek, P. Abraham, and J. E. Bowers, Self-consistent analysis of high-temperature effects on strained-layer multi-quantum well InGaAsP/InP lasers, *IEEE J. Quantum Electron.*, vol. 36, pp. 366–374, 2000.

[165] M. Grupen and K. Hess, Simulation of carrier transport and nonlinearities in quantum-well laser diodes, *IEEE J. Quantum Electron.*, vol. 34, pp. 120–140, 1998.

[166] G. J. Letal, J. G. Simmons, J. D. Evans, and G. P. Li, Determination of active-region leakage currents in ridge-waveguide strained-layer quantum-well lasers by varying the ridge width, *IEEE J. Quantum Electron.*, vol. 34, pp. 512–518, 1998.

[167] J. Piprek, D. I. Babic, and J. E. Bowers, Simulation and analysis of 1.55-micron double-fused vertical-cavity lasers, *J. Appl. Phys.*, vol. 81, pp. 3382–3390, 1997.

[168] Y. Zou, J. Osinski, P. Grodzinski, P. Dapkus, W. Rideout, W. Sharfin, J. Schlafer, and F. Crawford, Experimental study of Auger recombination, gain, and temperature sensitivity of 1.5 μm compressively strained semiconductor lasers., *IEEE J. Quantum Electron.*, vol. 29, pp. 1565–1575, 1993.

[169] G. Fuchs, C. Schiedel, A. Hangleiter, V. Harle, and F. Scholz, Auger recombination in strained and unstrained InGaAs/InGaAsP multiple quantum well lasers, *Appl. Phys. Lett.*, vol. 62, pp. 396–398, 1993.

[170] S. Seki, W.W.Lui, and K. Yokoyama, Explanation for the temperature insensitivity of the Auger recombination rates in 1.55 µm InP-based strained-layer quantum-well lasers, *Appl. Phys. Lett.*, vol. 66, pp. 3093–3095, 1995.

[171] H. C. Casey and P. L. Panish, Variation of intervalence band absorption with hole concentration in p-type InP, *Appl. Phys. Lett.*, vol. 44, pp. 82–84, 1984.

[172] C. H. Henry, R. A. Logan, F.R.Merritt, and J. P. Luongo, The effect of intervalence band absorption on the thermal behavior of InGaAsP lasers, *IEEE J. Quantum Electron.*, vol. 19, pp. 947–952, 1993.

[173] I. Joindot and J.L.Beylat, Intervalence band absorption coefficient measurements in bulk layer, strained and unstrained multiquantum well 1.55 µm semiconductor lasers, *Electron. Lett.*, vol. 29, pp. 604–606, 1993.

[174] G. Fuchs, J. Hoerer, A. Hangleiter, V. Haerle, F. Scholz, R.W.Glew, and L. Goldstein, Intervalence band absorption in strained and unstrained InGaAs multiple quantum well structures, *Appl. Phys. Lett.*, vol. 60, pp. 231–233, 1992.

[175] W.S.Ring, Examinationof intervalence band absorption and its reduction by strain in 1.55-µm compressively strained InGaAs/InP laser diodes, *Electron. Lett.*, vol. 30, pp. 306–308, 1994.

[176] T. Cho, H. Kim, Y. Kwon, and S. Hong, Theoretical study on intervalence band absorption in InP-based quantum-well laser structures, *Appl. Phys. Lett.*, vol. 68, pp. 2183–2185, 1996.

[177] R. F. Kazarinov and M. R. Pinto, Carrier transport in laser heterostructures, *IEEE J. Quantum Electron.*, vol. 30, pp. 49–53, 1994.

[178] J. Braithwaite, M. Silver, V. A. Wilkinson, E. P. O'Reilly, and A. R. Adams, Role of radiative and non-radiative processes on the temperature sensitivity of strained and unstrained 1.5 µm InGaAs(P) quantum well lasers, *Appl. Phys. Lett.*, vol. 67, pp. 3546–3548, 1995.

[179] L. J. P. Ketelsen and R. F. Kazarinov, Carrier loss in InGaAsP-InP lasers grown by hydride CVD, *IEEE J. Quantum Electron.*, vol. 34, pp. 811–813, 1995.

[180] Y. Yoshida, H. Watanabe, K. Shibata, A. Takemoto, and H. Higuchi, Analysis of characteristic temperature for InGaAsP BH lasers with p-n-p-n blocking layers using two-dimensional device simulator, *IEEE J. Quantum Electron.*, vol. 34, pp. 1257–1262, 1998.

[181] V. Mikhaelashvili, N. Tessler, R. Nagar, G. Eisenstein, A. G. Dentai, S. Chandrasakhar, and C. H. Joyner, Temperature dependent loss and overflow effects in quantum well lasers, *IEEE Photon. Technol. Lett.*, vol. 6, pp. 1293–1296, 1994.

[182] A. A. Bernussi, H. Temkin, D. Coblentz, and R. A. Logan, Effect of barrier recombination on the high temperature performance of quaternary multiquantum well lasers, *Appl. Phys. Lett.*, vol. 66, pp. 67–69, 1995.

[183] D. A. Ackerman, G. E. Shtengel, M. S. Hybertsen, P. A. Morgan, R. F. Kazarinov, T. Tanbun-Ek, and R. A. Logan, Analysis of gain in determining T_0 in 1.3 μm semiconductor lasers, *IEEE J. Select. Topics Quantum Electron.*, vol. 1, pp. 250–262, 1995.

[184] T. E. Sale, *Vertical Cavity Surface Emitting Lasers.* New York: Wiley, 1995.

[185] C. Wilmsen, H. Temkin, and L. A. Coldren, eds., *Vertical-Cavity Surface-Emitting Lasers.* Cambridge, UK: Cambridge Univ. Press, 1999.

[186] D. I. Babic, Y. Chung, N. Dagli, and J. E. Bowers, Modal reflectivity of quarter-wave mirrors in vertical-cavity lasers, *IEEE J. Quantum Electron.*, vol. 6, pp. 1950–1962, 1993.

[187] A. Karim, S. Bjorlin, J. Piprek, and J. Bowers, Long-wavelength vertical-cavity lasers and amplifiers, *IEEE J. Select. Topics Quantum Electron.*, vol. 6, pp. 1244–1253, 2000.

[188] V. Ustinov and A. Zhukov, GaAs-based long-wavelength lasers, *Semicond. Sci. Technol.*, vol. 15, pp. 41–54, 2000.

[189] A. Black, A. Hawkins, N. Margalit, D. Babic, J. Holmes, A.L., Y.-L. Chang, P. Abraham, J. Bowers, and E. Hu, Wafer fusion: materials issues and device results, *IEEE J. Select. Topics Quantum Electron.*, vol. 3, pp. 943–951, 1997.

[190] V. Jayaraman, T. Goodnough, T. Beam, F. Ahedo, and R. Maurice, Continuous-wave operation of single-transverse-mode 1310-nm VCSELs up to 115C, *IEEE Photon. Technol. Lett.*, vol. 12, pp. 1595–1597, 2000.

[191] A. Karim, J. Piprek, P. Abraham, D. Lofgreen, Y.-J. Chiu, and J. E. Bowers, 1.55-micron vertical-cavity laser arrays for wavelength-division multiplexing, *IEEE J. Select. Topics Quantum Electron.*, vol. 7, pp. 178–183, 2001.

[192] N. M. Margalit, J. Piprek, S. Zhang, D. I. Babić, K. Streubel, R. Mirin, J. R. Wesselmann, J. E. Bowers, and E. L. Hu, 64C continuous-wave operation of 1.5-micron vertical-cavity lasers, *IEEE J. Select. Topics Quantum Electron.*, vol. 3, pp. 359–365, 1997.

[193] M. G. Peters, B. J. Thibeault, D. B. Young, A. C. Gossard, and L. A. Coldren, Growth of beryllium doped AlGaAs/GaAs mirrors for vertical-cavity surface-emitting lasers, *J. Vac. Sci. Technol.*, vol. 12, pp. 3075–3083, 1994.

[194] J. Piprek, Y. A. Akulova, D. I. Babic, L. A. Coldren, and J. E. Bowers, Minimum temperature sensitivity of 1.55-micron vertical-cavity lasers at -30 nm gain offset, *Appl. Phys. Lett.*, vol. 72, no. 15, pp. 1814–1816, 1998.

[195] Y. M. Zhang, J. Piprek, N.Margalit, M.Anzlowar, and J. Bowers, Cryogenic performance of double-fused 1.5-micron vertical-cavity lasers, *IEEE J. Lightwave Technol.*, vol. 17, pp. 503–508, 1999.

[196] D. Babic, J. Piprek, K. Streubel, R. Mirin, N. Margalit, D. Mars, J. Bowers, and E. Hu, Design and analysis of double-fused 1.55 μm vertical-cavity lasers, *IEEE J. Quantum Electron.*, vol. 33, pp. 1369–1383, 1997.

[197] F. Salomonsson, K. Streubel, J. Bentell, M. Hammar, D. Keiper, R. Westphalen, J. Piprek, L. Sagalowicz, A. Roudra, and J. Behrend, Wafer fused p-InP/p-GaAs heterojunctions, *J. Appl. Phys.*, vol. 83, pp. 768–774, 1998.

[198] J. Piprek, Electro-thermal analysis of oxide-confined vertical-cavity lasers, *phys. stat. sol. (a)*, vol. 188, pp. 905–912, 2001.

[199] K. Black, P. Abraham, N. Margalit, E. Hegblom, Y.-J. Chiu, J. Piprek, J. Bowers, and E. Hu, Double-fused 1.5-micron vertical cavity lasers with record high To of 132K at room temperature, *Electron. Lett.*, vol. 34, pp. 1947–1949, 1998.

[200] D. I. Babic, *Double-Fused Long-Wavelength Vertical-Cavity Lasers*. PhD thesis, University of California at Santa Barbara, 1995.

[201] W. Both and J. Piprek, Thermal resistance of ridge-waveguide laser diodes based on GaAs, GaSb or InP, *J. Thermal Analys.*, vol. 37, pp. 61–71, 1991.

[202] J. Piprek, T. Troger, B. Schroter, J. Kolodzey, and C. S. Ih, Thermal conductivity reduction in GaAs-AlAs distributed Bragg reflectors, *IEEE Photon. Technol. Lett.*, vol. 10, no. 1, pp. 81–83, 1998.

[203] G. Chen, C. L. Tien, X. Wu, and J. S. Smith, Thermal diffusivity measurement of GaAs/AlGaAs thin-film structures, *ASME J. Heat Transfer*, vol. 116, pp. 325–331, 1994.

[204] N. M. Margalit, D. I. Babić, K. Streubel, R. Mirin, R. L. Naone, J. E. Bowers, and E. L. Hu, Submilliamp long wavelength vertical cavity lasers, *Electron. Lett.*, vol. 32, pp. 1675–1677, 1996.

[205] P. Bienstman, R. R. Baets, J. Vukusic, A. Larsson, M. Noble, M. Brunner, K. Gulden, P. Debernardi, L. Fratta, G. Bava, H. Wenzel, B. Klein, O. Conradi, R. Pregla, S. Riyopoulos, J.-F. Seurin, and S. L. Chuang, Comparison of optical VCSEL models of the simulation of oxide-confined devices, *IEEE J. Quantum Electron.*, vol. 37, pp. 1618 –1631, 2001.

[206] M. J. Noble, J. P. Loehr, and J. A. Lott, Analysis of microcavity VCSEL lasing modes using a full vector weighted index method, *IEEE J. Quantum Electron.*, vol. 34, pp. 1890–1903, 1998.

[207] D. Burak, J. Moloney, and R. Binder, Macroscopic versus microscopic description of polarization properties of optically anisotropic vertical-cavity surface-emitting lasers, *IEEE J. Quantum Electron.*, vol. 36, pp. 956–970, 2000.

[208] G. P. Bava, P. Debernardi, and L. Fratta, Three-dimensional model for vectorial fields in vertical-cavity surface-emitting lasers, *Phys. Rev. A*, vol. 63, p. 023816, 2001.

[209] G. Hadley, K. Lear, M. Warren, K. Choquette, J. Scott, and S. Corzine, Comprehensive numerical modeling of vertical-cavity surface-emitting lasers, *IEEE J. Quantum Electron.*, vol. 32, pp. 607–616, 1996.

[210] J. Piprek, H. Wenzel, and G. Sztefka, Modeling thermal effects on the light vs. current characteristic of gain-guided vertical-cavity surface-emitting lasers, *IEEE Photon. Technol. Lett.*, vol. 6, pp. 139–142, 1994.

[211] B. J. Thibeault, T. A. Strand, T. Wipiejewski, M. G. Peters, D. B. Young, S. Corzine, L. A. Coldren, and J. Scott, Evaluating the effects of optical and carrier losses in etched-post vertical cavity lasers., *J. Appl. Phys.*, vol. 78, pp. 5871–5875, 1995.

[212] D. Young, J. Scott, F. Peters, M. Peters, M. Majewski, B. Thibeault, S. Corzine, and L. Coldren, Enhanced performance of offset-gain high-barrier vertical-cavity surface-emitting lasers, *IEEE J. Quantum Electron.*, vol. 29, pp. 2013–2022, 1993.

[213] B. Tell, K. Brown-Goebeler, R. Leibenguth, F. Baez, and Y. Lee, Temperature dependence of GaAs-AlGaAs vertical cavity surface emitting lasers, *Appl. Phys. Lett.*, vol. 60, pp. 683–685, 1992.

[214] B. Lu, P. Zhou, J. Cheng, K. Malloy, and J. Zolper, High temperature pulsed and continuous-wave operation and thermally stable threshold characteristics of vertical-cavity surface-emitting lasers grown by metalorganic chemical vapor deposition, *Appl. Phys. Lett.*, vol. 65, pp. 1337–1339, 1994.

[215] K. Streubel, S. Rapp, J. Andre, and J. Wallin, Room-temperature pulsed operation of 1.5 μm vertical cavity lasers with an InP-based Bragg reflector, *IEEE Photon. Technol. Lett.*, vol. 8, pp. 1221–1223, 1996.

[216] S. Rapp, J. Piprek, K. Streubel, J. Andre, and J. Wallin, Temperature sensitivity of 1.54-micron vertical-cavity lasers with an InP-based Bragg reflector, *IEEE J. Quantum Electron.*, vol. 33, no. 10, pp. 1839–1845, 1997.

[217] S. Nakamura, S. Pearton, and G. Fasol, *The Blue Laser Diode*. Berlin: Springer-Verlag, 2000.

[218] W. Götz, N. M. Johnson, J. Walker, D. P. Bour, and R. A. Street, Activation of acceptors in Mg-doped GaN grown by metalorganic chemical vapor deposition, *Appl. Phys. Lett.*, vol. 68, pp. 667–669, 1996.

[219] W. Götz, N. M. Johnson, C. Chen, H. Liu, C. Kuo, and W. Imler, Activation energies of Si donors in GaN, *Appl. Phys. Lett.*, vol. 68, pp. 3144–3146, 1996.

[220] S. Nakamura, T. Mukai, and M. Senoh, In situ monitoring and Hall measurements of GaN grown with GaN buffer layers, *J. Appl. Phys.*, vol. 71, pp. 5543–5549, 1992.

[221] J. D. Albrecht, R. P. Wang, P. P. Ruden, M. Farahmand, and K. F. Brennan, Electron transport characteristics of GaN for high temperature device modeling, *J. Appl. Phys.*, vol. 83, pp. 4777–4781, 1998.

[222] S. Nakamura, M. Senoh, and T. Mukai, Highly p-type Mg-doped GaN films grown with GaN buffer layers, *Japan. J. Appl. Phys., Part 2*, vol. 30, pp. L1708–L1711, 1991.

[223] M. Rubin, N. Newman, J. S. Chan, T. C. Fu, and J. T. Ross, p-type gallium nitride by reactive ion-beam molecular beam epitaxy with ion implantation, diffusion, or coevaporation of Mg, *Appl. Phys. Lett.*, vol. 64, pp. 64–66, 1994.

[224] J. F. Muth, J. H. Lee, I. K. Shmagin, R. M. Kolbas, H. C. Casey, B. P. Keller, U. K. Mishra, and S. P. DenBaars, Absorption coefficient, energy gap, exciton binding energy, and recombination lifetime of GaN obtained from transmission measurements, *Appl. Phys. Lett.*, vol. 71, pp. 2572–2574, 1997.

[225] A. V. Dmitriev and A. L. Oruzheinikov, The rate of radiative recombination in the nitride semiconductors and alloys, *MRS Internet J. Nitride Semicond. Res.*, vol. 1, p. 46, 1996.

[226] S. R. Lee, A. F. Wright, M. H. Crawford, G. A. Petersen, J. Han, and R. M. Biefeld, The band-gap bowing of $Al_xGa_{1-x}N$ alloys, *Appl. Phys. Lett.*, vol. 74, pp. 3344–3346, 1999.

[227] A. C. Abare, *Growth and Fabrication of Nitride-Based Distributed Feedback Laser Diodes*. PhD thesis, University of California at Santa Barbara, 2000.

[228] Y. C. Yeo, T. C. Chong, M. F. Li, and W. J. Fan, Analysis of optical gain and threshold current density of wurtzite InGaN/GaN/AlGaN quantum well lasers, *J. Appl. Phys.*, vol. 83, pp. 1813–1819, 1998.

[229] Y. C. Yeo, T. C. Chong, and M. F. Li, Uniaxial strain effect on the electronic and optical properties of wurtzite GaN-AlGaN quantum-well lasers, *IEEE J. Quantum Electron.*, vol. 34, pp. 2224–2232, 1998.

[230] S. H. Park and S. L. Chuang, Many-body optical gain of wurtzite GaN-based quantum-well lasers and comparison with experiment, *Appl. Phys. Lett.*, vol. 72, pp. 287–289, 1998.

[231] F. Bernardini, V. Fiorentini, and D. Vanderbilt, Spontaneous polarization and piezoelectric constants of III-V nitrides, *Phys. Rev. B*, vol. 56, pp. R10024–R10027, 1997.

[232] F. Bernardini, V. Fiorentini, and D. Vanderbilt, Accurate calculation of polarization-related quantities in semiconductors, *Phys. Rev. B*, vol. 63, p. 193201, 2001.

[233] F. Bernardini and V. Fiorentini, Nonlinear macroscopic polarization in iii-v nitride alloys, *Phys. Rev. B*, vol. 64, p. 085207, 2001.

[234] L. H. Peng, C. W. Chuang, and L. H. Lou, Piezoelectric effects in the optical properties of strained InGaN quantum wells, *Appl. Phys. Lett.*, vol. 74, pp. 795–797, 1999.

[235] F. Della Sala, A. Di Carlo, P. Lugli, F. Bernardini, V. Fiorentini, R. Scholz, and J.-M. Jancu, Free-carrier screening of polarization fields in wurtzite

GaN/InGaN laser structures, *Appl. Phys. Lett.*, vol. 74, pp. 2002–2004, 1999.

[236] C. A. Flory and G. Hasnain, Modeling of GaN optoelectronic devices and strain-induced piezoelectric effects, *IEEE J. Quantum Electron.*, vol. 37, pp. 244–253, 2001.

[237] M. J. Bergmann and H. C. Casey, Optical-field caalculations for lossy multiple-layer AlGaN/InGaN laser diodes, *J. Appl. Phys.*, vol. 84, pp. 1196–1203, 1998.

[238] M. M. Y. Leung, A. B. Djurisic, and E. H. Li, Refractive index of InGaN/GaN quantum wells, *J. Appl. Phys.*, vol. 84, pp. 6312–6317, 1998.

[239] S. Nakamura, M. Senoh, N. Iwasa, and S. Nagahama, High-power InGaN single-quantum-well-structure blue and violet light-emitting diodes, *Appl. Phys. Lett.*, vol. 67, pp. 1868–1870, 1995.

[240] S. D. Lester, F. A. Ponce, M. G. Craford, and D. A. Steigerwald, High dislocation densities in high efficiency GaN-based light-emitting diodes, *Appl. Phys. Lett.*, vol. 66, pp. 1249–1251, 1995.

[241] S. Nakamura, Development and future prospects of InGaN-based LEDs and LDs, in *Introduction to Nitride Semiconductor Blue Lasers and Light-Emitting Diodes* (S. Nakamura and S. F. Chichibu, eds.), pp. 317–350, London: Taylor & Francis, 2000.

[242] S. F. Chichibu, Y. Kawakami, and T. Sota, Emission mechanisms and excitons in GaN and InGaN bulk and QWs, in *Introduction to Nitride Semiconductor Blue Lasers and Light-Emitting Diodes* (S. Nakamura and S. F. Chichibu, eds.), pp. 153–270, London: Taylor & Francis, 2000.

[243] K. P. O'Donnell, T. Breitkopf, H. Kalt, W. van der Stricht, I. Moerman, P. Demeester, and P. G. Middleton, Optical linewidhts of InGaN light emitting diodes and epilayers, *Appl. Phys. Lett.*, vol. 70, pp. 1843–1845, 1997.

[244] P. Fischer, J. Christen, and S. Nakamura, Spectral electroluminescence mapping of a blue InGaN single quantum well light-emitting diode, *Japan. J. Appl. Phys., Part 2*, vol. 39, pp. L129–L132, 2000.

[245] M. Osinski and D. L. Barton, Life testing and degradation mechanisms in InGaN LEDs, in *Introduction to Nitride Semiconductor Blue Lasers and Light-Emitting Diodes* (S. Nakamura and S. F. Chichibu, eds.), pp. 271–315, London: Taylor & Francis, 2000.

[246] S. J. Lee, Analysis of InGaN high-brightness light-emitting diodes, *Japan. J. Appl. Phys., Part 1*, vol. 37, pp. 5990–5993, 1998.

[247] S. Nakamura, M. Senoh, S. I. Nagahama, N. Iwasa, T. Yamada, T. Matsushita, H. Kiyoku, Y. Sugimoto, T. Kozaki, I. Umemoto, M. Sano, and K. Chocho, Violet InGaN/GaN/AlGaN-based laser diodes with an output power of 420 mW, *Japan. J. Appl. Phys., Part 2*, vol. 37, pp. L627–L629, 1998.

[248] W. W. Chow and S. W. Koch, Theory of laser gain in group-III nitride quantum wells, in *GaN and Related Materials* (S. J. Pearton, ed.), pp. 235–262, New York: Gordon & Breach, 1999.

[249] J. Piprek, K. White, and A. SpringThorpe, What limits the maximum ouptut power of long-wavelength AlInGaAs laser diodes ?, *IEEE J. Quantum Electron.*, vol. 38, pp. 1253–1259, 2002.

[250] M. Kneissl, W. S. Wong, D. W. Treat, M. Teepe, N. Miyashita, and N. M. Johnson, Continuous-wave operation of InGaN multiple-quantum-well laser diodes on copper substrates obtained by laser liftoff, *IEEE J. Select. Topics Quantum Electron.*, vol. 7, pp. 188–191, 2001.

[251] S. Z. Zhang, Y.-J. Chiu, P. Abraham, and J. E. Bowers, 25-GHz polarization-insensitive electroabsorption modulators with traveling-wave electrodes, *IEEE Photon. Technol. Lett.*, vol. 11, pp. 191–193, 1999.

[252] S. Z. Zhang, *Traveling-Wave Electroabsorption Modulators*. PhD thesis, University of California at Santa Barbara, 1999.

[253] J. Piprek, Y.-J. Chiu, J. E. Bowers, C. Prott, and H. Hillmer, High-efficiency multi-quantum well electroabsorption modulators, in *Integrated Optoelectronics*, vol. 2002-4, (Pennington), pp. 139–149, ECS - The Electrochem. Soc., 2002.

[254] G. Agrawal, *Fiber-Optic Communication Systems*. Singapore: Wiley, 1993.

[255] D. Lasaosa, Y. J. Chiu, J. Piprek, and J. E. Bowers, Traveling-wave amplification photodetector (TAP detector), in *13th Lasers and Electro-Optics Society Annual Meeting*, (Piscataway), pp. 260–261, Institute of Electrical and Electronic Engineers, 2000.

[256] A. W. Snyder and J. D. Love, *Optical Waveguide Theory*. London: Chapmann and Hall, 1983.

Index

Γ point, 14

Auger recombination heat, 145
Axial approximation, 27

A

ABCD matrix, 117
Absorbance, 102
Absorption, 71, 85, 87, 227
Absorption coefficient, 72, 86, 89, 102, 158, 161
Absorption spectrum, 215
Absorption strength, 214
Absorption, band-to-band, 88, 121, 213, 215
Absorption, cavity length dependence, 161
Absorption, edge, 215
Absorption, electro-, 214
Absorption, free-carrier, 89, 158
Absorption, interconduction valley, 90
Absorption, intervalence band, 90, 160
Absorption, intraband, 90
Absorption, modal, 167, 221
Absorption, of TE and TM mode, 214
Absorption, quantum-well, 214
Absorption, residual, 220
Absorption, reststrahlen, 91
Absorption, temperature dependence, 168
Alloy scattering, 143
Amplifier, 227
Asada scattering model, 132
Astigmatic beam, 116
Auger recombination, 68, 157, 163, 165

B

Band diagram, 52, 153, 175, 193, 204, 218, 230
Band gap, 8, 20, 22, 23
Band gap bowing, 43
Band gap narrowing, 22
Band gap renormalization, 22, 135, 158, 192
Band gap wavelength, 3, 94
Band gap, carrier concentration dependence, 22
Band gap, nitride alloys, 191
Band gap, temperature dependence, 21, 158, 191
Band offset, 43, 134, 230
Band structure, 16
Band structure, $k \cdot \vec{p}$ method, 23
Band structure, BULK GaAs, 17
Band structure, bulk GaAs, 29
Band structure, bulk GaN, 37
Band structure, four-band model, 32
Band structure, GaAs quantum well, 41
Band structure, InGaN quantum well, 191
Band structure, Luttinger–Kohn model, 31
Band structure, strain effects, 131
Band structure, three-band model, 31
Band structure, two-band model, 24
Band structure, wurtzite, 32

Printed and bound by CPI Group (UK) Ltd, Croydon, CR0 4YY

12/05/2025

01866873-0002